NEW YORK INSTITUTE OF FINANCE

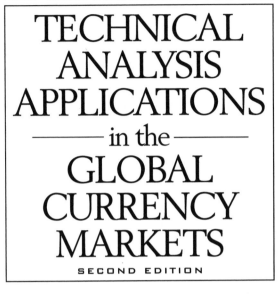

TECHNICAL ANALYSIS APPLICATIONS

— in the —

GLOBAL CURRENCY MARKETS

SECOND EDITION

NEW YORK INSTITUTE OF FINANCE

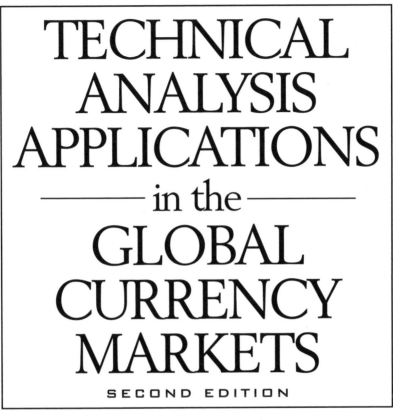

TECHNICAL ANALYSIS APPLICATIONS
— in the —
GLOBAL CURRENCY MARKETS

SECOND EDITION

CORNELIUS LUCA

NEW YORK INSTITUTE OF FINANCE
NEW YORK • TORONTO • SYDNEY • TOKYO • SINGAPORE

Library of Congress Cataloging-in-Publication Data

Luca, Cornelius.
   Technical analysis applications in the global currency markets /
Cornelius Luca.—2nd ed.
     p. cm.
   Includes bibliographical references and index.
   ISBN 0-7352-0147-1 (cloth)
    1. Foreign exchange futures—Forecasting. 2. Investment analysis
   I. Title.
   HG3853.L83 2000
   332.4'5—dc21   00-020841
                                               CIP

This publication is designed to provide accurate and authoritative information in regard to the subject matter covered. It is sold with the understanding that the publisher is not engaged in rendering legal, accounting, or other professional service. If legal advice or other expert assistance is required, the services of a competent professional person should be sought.

. . . From a Declaration of Principles Jointly Adapted by a Committee of the American Bar Association and a Committee of Publishers and Associations.

Printed in the United States of America

10 9 8 7 6 5 4 3 2 1

ISBN 0-7352-0147-1

 NEW YORK INSTITUTE OF FINANCE
An Imprint of Prentice Hall Press
Paramus, NJ 07652

Visit us at www.phdirect.com/business

NYIF and NEW YORK INSTITUTE OF FINANCE are trademarks of Executive Tax Reports, Inc. used under license by Prentice Hall Direct, Inc.

*To my wife, Sandra, and my daughter, Gwendolyn,
for all the help and support*

Contents

PART 2
TYPES OF CHARTS / 107

PART 3
QUANTITATIVE METHODS OF ANALYSIS / 209

PREFACE

For the first time after taking off in earnest in 1973, the foreign exchange industry experienced two major changes in the late 1990s that completely altered its outlook, necessitating the revision of this book: the introduction of the euro and the expansion of electronic trading.

After intense and prolonged negotiations, the European Monetary Union launched the euro at the beginning of 1999 in an all-out effort not only to create a common currency for its members, but also to develop the only basket currency that might dethrone the dollar from its worldwide dominating position.

While electronic trading has been an integral part of currency trading from the very beginning, the interbank activity used to be fairly equally divided between voice, or human brokers, and electronic means. The introduction of electronic brokers in 1992 had such an impact on trading that most voice brokers were forced to either merge or fold by the end of the millennium.

The oldest of the financial markets, foreign exchange is also the newest in contemporary times. Its inherent profitability and the limited credit risk are but two aspects that attracted a flock of traders looking to speculate or hedge their exposures. As the demand for trading currencies ballooned, so did the competition. The maturing of the market makes the way to easy money much more convoluted.

Technical analysis can give you an edge in this fast-moving, potentially lucrative market. Based on the general premise that history will repeat itself if you recognize the patterns, studying the charts provides you with quantifiable and flexible methods of forecasting.

What is a chart? A chart is the most comprehensive and colorful map of the traders' souls. We have all heard—and wanted to believe—the beautiful war stories of the traders who never lost money, who always took their profits before damaging events occurred, or who never missed joining a trend at the very moment of its inception. Once you have learned to use technical analysis to your advantage, those stories can be about you.

You get closer to the truth when you study charts. Only on them are you able to spot the killing fields of false breakouts, feel the fresh air of a major trend, fight the shifting sands of a wealth-sapping sideways move, or score big on a trend reversal formation that you were able to identify a little ahead of the crowd.

Currency trading is very much like a war, a complex one that combines head-on attacks and guerrilla fights. You will know some of your foes, but many will come without a face. Throughout all this, a fundamental question arises: Are you the general plotting the war and choosing the winning strategy? Or are you the soldier pushed to the front and fighting desperately in a bloody struggle that is not yours?

It really is your choice whether to be the soldier or the general. Charts may be there for all to see, but not everyone understands or knows how to read the signs. This book gives you all the tactics and weapons you need to fight each and every one of your battles. You can decide what is best for your personality and your specific trading style. And the next time you stand on the edge of the battlefield, you will be able to confidently select and implement a course of action, not just charge in hoping for the best.

This unique resource for players in the foreign exchange markets, *Technical Analysis Applications in the Global Currency Markets,* strives to answer questions that benefit aspiring and seasoned traders, speculators and hedgers, aggressive and conservative investors, and large and small players alike. These questions include:

- Can you really get any trading information from charting?
- Why is this information valuable?
- What is the significance of the trends?
- What are the major patterns and how can you take advantage of them?
- How can you read more trading signals in point and figure charts and candlestick charts?

- What are all the quantitative methods and what do they mean?
- Can you really take advantage of the Gann methods and the Elliott Wave principle in day-to-day trading?

This book is divided into three parts:

Part 1: Presents the importance of technical analysis in currency forecasting, introduces you to the types of charts, and analyzes the trends and the major trend formations.

Part 2: Analyzes all types of charts and details their unique characteristics so that you can enhance your trading.

Part 3: Takes you one step beyond simple chart reading and focuses in a user-friendly manner on moving averages, oscillators, and an array of other indicators that filter out many false signals, while generating valid trading levels.

Leading information services have graciously contributed the best of their products. You can see what the top players look at, compare different systems, and test your grasp of currency charting in the software program provided by Strategem Software International, Inc.

While there are no perfect answers for everyone, or for every occasion, a thorough application of technical analysis in your currency trading is a must if you really want to go to the top. Do not go for fads, but test and find the set of methods that you find helpful.

Good luck and bountiful hunting!

Cornelius Luca

ACKNOWLEDGMENTS

Many thanks to everyone at the New York Institute of Finance for making this book possible. Specifically, I want to extend my gratitude for the help and support to Robert Gulick, President, Dana Orenstein, Director, and William A. Rini and Paul McQuarry, former directors.

The help and support of Ellen Schneid Coleman, from Prentice Hall Press and Fred Dahl, of Inkwell Publishing Services, were instrumental for the successful completion of the book.

The charts in this training manual have been made possible by the following:

Michael Cuttone, Vice President, Workstation & Technical Analysis Products, Bridge Information Systems, Inc.

Elizabeth DeMarse, Director of Marketing, Bloomberg Financial Markets

Michael DuVally, Managing Editor, Financial Markets, Bridge Information Systems, Inc.

Robert Hafer, Director of Financial Publishing, Bridge Information Systems, Inc.

Paul Lowe, Executive Editor, Bridge Information Systems, Inc.

Nick Van Nice, Publisher, Commodity Trend Service

Darril Malloy, Manager, Bloomberg Financial Markets

Stephen Onstad, National Marketing Representative & New York Regional Manager, CQG

Michele Orton, Marketing/Advertising Assistant, CQG

Kevin Pendley, Chicago Bureau Chief, Bridge Information Systems, Inc.

John Schillaci, COO, FutureSource

Kevin McDermott, Account Executive, FutureSource

Andrew White, VP Open Systems, Reuters

Vicki Brown, Manager, Reuters

James Ritter, President, Stratagem Software International, Inc.

Saeed Shamsnia, CEO, Stratagem Software International, Inc.

PART 1

TRENDS AND MAJOR CHART FORMATIONS

chapter one

*F*UNDAMENTAL PRINCIPLES OF TECHNICAL ANALYSIS

Currency traders study the past behavior of currency prices on charts in order to forecast their future performance. Although the use of chart analysis has increased significantly since the mid-1980s, compliments of technological breakthroughs, such analysis is nothing new, having been used increasingly over the last two centuries. European grains traders first started to plot price levels in the sixteenth century. Seiki Shimizu, a famous Japanese commodities trader and the author of *The Japanese Chart of Charts* (Tokyo Futures Trading Publishing Co., Tokyo, 1986), a comprehensive study of candlestick charting, estimates that the beginning of the rice market during the 1700s quickly triggered the use of the first charts in Japan. Currency traders had to wait a while longer, as modern forex trading only started after 1971, following the demise of the pegged currency system of the Bretton Woods Accord. Today, technical analysis has emerged as the most powerful forecasting tool for the fast and complex world of foreign exchange.

Commodity traders started to analyze past price behavior because they felt that commodity prices reflect the action of all available factors, that is, all the changes in the balance between supply and demand as caused by each and every trader's reaction to economic, political or psychological changes. Later, observations on the seasonality of crops evolved into the idea of historical repetition in the market. In time, traders observed other patterns on their charts. Finally, the empirical evidence showed that, once the market starts moving, its direction and momentum are likely to continue for a while, before new and stronger forces are able to change the balance between buyers and sellers, and turn the tide.

Thus three factors slowly crystallized in traders' minds as the basis of chart analysis:

1. *Price,* the ultimate result of all the market forces.
2. *Repetition of the market price behavior.*
3. *The market tendency to move in trends.*

Since the identification of these factors, there has been a vigorous and continuous effort to refine the process of technical analysis in order to profitably forecast future market behavior.

Technical Analysis on Balance

To the untrained eye, technical analysis may seem, at times, confusing. On one hand, everybody seems to be talking about it, using it and profiting from it. On the other hand, a handful of academicians compare it with the structure of a house of cards.

Where is the truth?

Pros . . .

The main strengths of technical analysis are its flexibility, its flexibility and its flexibility.

First, there is flexibility with regard to the *underlying instrument.* Traders who specialize in one currency, such as the Swiss franc, may easily apply their technical expertise to trading another currency, such as the Japanese yen. Traders do this when the trading activity in their specialty currency temporarily slows down.

Second, there is flexibility regarding the *markets.* Traders who specialize in spot trading can make a smooth transition to dealing currency futures or cross currencies by using charts, because the same technical principles apply regardless of the market. Different players have different trading styles and objectives. However, whether to speculate or hedge, you use the same charts and chart studies.

Finally, there is flexibility regarding the *time frame.* Most currency traders are day traders. However, your analysis has the same general basis if you take long-term strategic positions.

Besides its flexibility, technical analysis has a "user-friendly" feature. The computations are easy, and the technical services are becoming increasingly sophisticated and reasonably priced. Currently, the market enjoys a multitude of financial information services that offer printed and electronic charts. Technically inclined traders have to focus only on interpreting them, not on producing them.

Technical analysis will enhance your trading confidence because it makes it easy to identify your entry and exit levels. You will stop guessing and truly start forecasting.

. . . And Cons

All good things get their share of criticism, and so does technical analysis. Criticism is healthy when it is constructive. Criticism is particularly good for the technical analysis of foreign exchange because it helps newcomers to better understand its characteristics and how it can enhance their trading performance.

The technical analysis naysayers usually focus on two general aspects:

1. The Random Walk Theory.
2. The self-fulfilling prophecy.

The Random Walk Theory

The Random Walk Theory, developed by Paul Cootner in 1964, is based on the efficient market hypothesis, which states that prices move randomly versus their intrinsic value. According to this theory, neither technical nor fundamental analyses can predict any future price change. The Random Walk Theory has three versions: the strong version, the semistrong version, and the weak version.

1. The *strong version* holds that it is impossible to forecast anything based on any information.
2. The *semistrong version* states that forecasting is impossible if it is based on publicly available information. Since most information in foreign exchange—along with the charts reflecting that information—are publicly available, by extrapolation, forecasting should generally be precluded. Exceptions are allowed in the case of proprietary information. A trading desk receiving an order to buy a very large amount of currency is therefore able to "forecast" the behavior of the market.

3. The *weak version* maintains that an efficient market with instanta-
    neous access to information will discount everything, so that the use
    of past data is useless for forecasting. Academicians and traders alike
    have labeled the foreign exchange, among all the financial markets,
    as the most efficient market. Unlike the equity markets, which trade
    tens of thousands of issues, most of the forex markets, which trade
    daily about $1.4 trillion, focus on about four major currency pairs:
    euro/US dollar, US dollar/Japanese yen, British pound/US dollar,
    and US dollar/Swiss franc (see Figure 1.1).

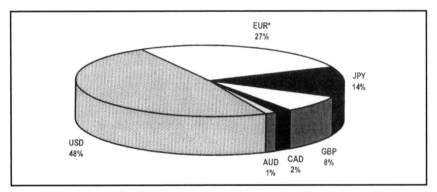

Figure 1.1. Market share of the foreign currencies as estimated after the introduction of
the euro. (*Source:* Bank For International Settlements)

In addition, as illustrated in Figure 1.2, in foreign exchange you
only have five instruments:

1. *Spot:* A deal that matures in two business days (with the exception
    of the Canadian dollar, which matures in one business day).
2. *Forward outright:* A deal that matures generally on a date past the
    spot date.
3. *Futures:* A special case of a forward outright.
4. *Currency swap:* A deal in which you simultaneously buy and sell, or
    sell and buy, the same amount of the same currency with the same
    counterparty, on different dates and at different rates.
5. *Options:* The right or the obligation to trade a currency in the future
    based on a predetermined amount at a predetermined rate

The majority of the market trades the spot, followed by swaps, forward outright, options, and futures (see Figure 1.3).

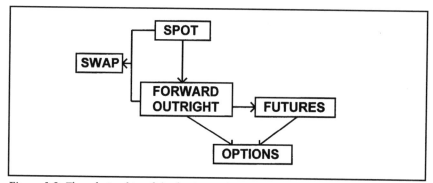

Figure 1.2. The relationship of the foreign exchange instruments.

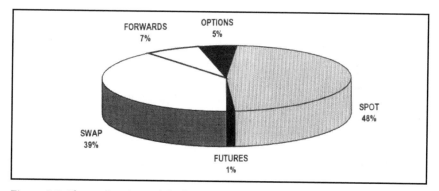

Figure 1.3. The market share of the foreign exchange instruments.
(*Source:* Bank For International Settlements)

If the hypothesis of market efficiency holds, there is no niche in which an individual trader can speculate at the expense of the rest of the market because all past information has already been reflected in the price.

Any "turf advantage" or any other type of "inside information" is instantly gathered and digested by the markets. Consequently, all market participants are precluded from consistently having any advantage over the rest of the market. The instantaneous access to information around the world should only add to the validity of the Random Walk Theory.

Is it really true? Naturally, all markets are random to some degree. The foreign exchange market, which has such a large volume that no individual player can have a consistent or long-term impact on the direction of a major currency, is perhaps more random than most. Indeed, until recently, the academic community has generally been unable to prove the existence of any particular patterns governing price activity.

However, academic studies, such as the tests conducted at both the University of California and the University of Wisconsin, which took advantage of unprecedented computer capability, fully contradict the Random Walk Theory. Even if all traders around the world receive exactly the same information at exactly the same time, they do not all trade in the same manner. Traders are so diverse that, given the same information, some will buy while others will sell. Also, some traders are long-term–oriented and more flexible with regard to the entry levels; thus they seem inefficient or "wrong" to short-term traders. Also, some traders have a so-called "turf advantage" or "inside information." This advantage derives from the orders that a large bank has from its corporate customers, which can be substantial on either an individual or a continuous basis. Traders having access to this type of information can outperform the rest of the market in either the short or medium term. Furthermore, corporations, in hedging their currency exposure, may enter seemingly inefficient positions. Other large players temporarily go against the market following an obvious chart signal, but just long enough to make a sizable profit. Finally, central banks get involved in the forex market to maintain good trading conditions and a healthy domestic economy, not to make a profit—a double-pronged goal that diverges from any rule of efficiency. With such a diversity of interests in an otherwise efficient market, it is rather difficult to see how the Random Walk Theory can realistically be applied.

The claim that chart patterns have no forecasting significance really makes no sense.

Take a look at the long-term dollar/yen chart in Figure 1.4. In addition to a clear down trend, you can also identify an interesting pattern: The dollar/yen spot rate moves in increments of 20 yen, or 2,000 pips in the long term. From its high at around 240, the currency dropped to 220, 200, 180, 160, 140, back up to around 160, and then down to 140 and 120. Every 2,000 pips, the US dollar/Japanese yen spot rate found a trading plateau. After reaching the 120 level the third time, the dollar/yen spot rate reached and broke the 100 barrier in June 1994, on its way to the all-time low of 79.75, or about 80.00 on April 19, 1995. This pattern is neither an exception nor a coincidence, and 18 consecutive trading years stand as the living proof. From the 80 level the dollar/yen bounced back to 100, 120, and then to 140. This unique characteristic, generated by the commercial impact of the huge interest of the *keiretsu,* the giant Japanese consortiums, plays a key role in identifying the level of the all-time high of the Japanese yen against the US dollar. It also tells you that, as long as the US dollar/Japanese yen holds above 100, the target is 120.

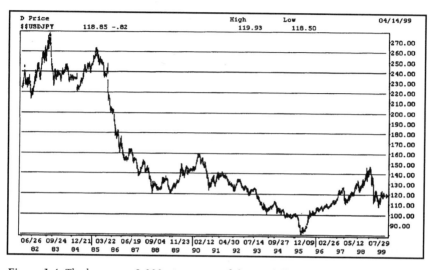

Figure 1.4. The long-term 2,000-pip pattern of the US dollar/Japanese yen market: 1981–1999. (*Source:* Bridge Information Systems, Inc.)

The Self-Fulfilling Prophecy

The *self-fulfilling prophecy* criticism has been around for a long time, and technical analysts wouldn't mind this theory—if it proved to be true. Trading would become easy: Just post the capital and have the broker/guru trade for you. Not only would it be the fastest way of getting rich overnight, but think of all the trader-sized egos that would really be touched.

The self-fulfilling prophecy, however, has too many holes to support anything but the most superficial criticism, because chart interpretation is very subjective and personal. It requires an individual, innate talent mixed with personally acquired and refined research and understanding. As discussed under the Random Walk Theory, since different traders have different objectives, time frames, and trading styles, it is difficult to predict whose prophecy will be fulfilled. Moreover, if the time-honored law of supply and demand does not provide the backing, any technical-related price overshooting will be shortlived.

Chartists are not trying to create an artificial environment in which financial instruments can be maneuvered at will. They cannot create trends and reversal formations. They do, however, try hard to identify any patterns or signals in the behavior of the currency in the existing environment so as to anticipate future price activity.

Is the Past a Guide to the Future?

This brings us to the question of whether past information can predict the future. It can. And you don't have to venture all the way to the foreign exchange world to see this, because using past information to build forecasting models is standard in most industries. As a simple example, you all have seen magazine ads depicting smiling couples on empty beaches in the Caribbean. These ads promise you the vacation of your life at a very low price in the month of September, just $300 for six nights! With no previous information at hand, the deal may look too good to pass up. However, you should have some additional information. Year after year, September has been the peak month of the hurricane season in the Caribbean, a fact that should make it easy to forecast the September weather for at least the next season.

Since currency trading in particular is not a game of darts, precedent is very important in helping a chartist forecast a currency's path into the future. It is the lack of historic information, not its availability, that is detrimental. And it may be the failure to understand the lessons of the past, rather than the past itself, that triggered the debate.

Technical analysis has been proved over and over again in the financial markets. The majority of currency traders use it extensively. Never has there been such a consistently useful tool as charting, nor one that was able to make such a powerful impact on traders' decision processes.

Despite all the technological breakthroughs, charting remains more of an art than a science. Yet the more refined traders become, the better is their forecasting performance.

While arguments continue, the technicians, in huge numbers, focus their expertise on forecasting the price activity in the future.

For those who accept that the past can be a guide to the future, let's move ahead with three important points in mind:

1. The price is a comprehensive reflection of all the market forces.
2. Price movements are historically repetitive.
3. Price movements are trend followers.

Dow Theory

Charles Dow is the preeminent technical analyst. His theory is the foundation of all consequent charting studies. His research, based solely on the stock market, was published at the end of the nineteenth century only as editorials in an upstart financial newspaper called *The Wall Street Journal*. Actually, he didn't name his empirical observations at all. The "Dow Theory" was coined later by technicians celebrating our number-one technician. We will present in a nutshell only five of the six topics Dow emphasized most, because they apply to foreign exchange:

1. *The market discounts everything.* At any given time, all market information and forces are reflected in the currency prices.
2. *The market has three trends: primary, secondary, and minor.* The secondary trend is a correction to the primary trend and may retrace one-third, one-half or two-thirds from the primary trend.
3. *The primary trend has three phases: accumulation, run-up/run-down, and distribution.* In the accumulation phase the shrewdest traders enter new positions. In the run-up/run-down phase, the majority of the market finally "sees" the move and jumps on the bandwagon. Finally, in the distribution phase, the keenest traders take their profits and close their positions while the general trading interest slows down in an overshooting market.
4. *Volume must confirm the trend.* This condition is a tough one for currency traders because, as you'll see in the latter part of this chapter, we just don't have accurate volume figures (with the exception of the currency futures market).
5. *Trends exist until their reversals are confirmed.* Figures 1.5 and 1.6 show you examples of reversals in a bullish and a bearish currency market. In Figure 1.5, the selling signals occur at points A and B when the currency falls below previous lows. In Figure 1.6, the buying signals occur at points A and B when the currency exceeds the previous highs.

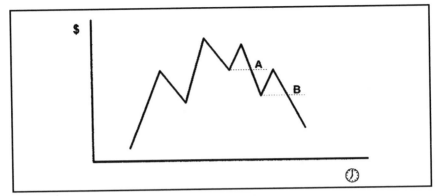

Figure 1.5. A reversal of a bullish currency.

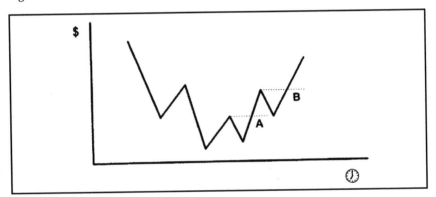

Figure 1.6. A reversal of a bearish currency.

A century later, the Dow Theory continues to be applied in a variety of ways. Some are simple, as in head and shoulders trend reversal formation; some more complex, such as in the Elliott Wave. Despite some shortcomings regarding the timeliness of the trading signals, studies conducted at the University of Wisconsin by Dr. William Brock and Dr. Blake LeBaron proved that the Dow Theory remains a reliable and profitable forecasting tool for technical analysts.

The Importance of Cycles

Cycles are the propensity for events to repeat themselves at roughly the same time. Cycle identification is a powerful tool that can be used in both the long and the short term. The longer the term, the more significance a cycle has. For instance, the weekly chart of the FINEX® US Dollar Index between 1974 and 1995 (see Figure 1.7) shows a long-term cycle of roughly seven years, with a low of 81.85 in September 1978, an all-time high of 164.72 in February 1985, and an all-time low of 78.19 in September 1992. Given more detail, you really have not seven-year intervals, but rather five years and five months in the first period and seven years and seven months in the latter. It is impossible to gather enough currency prices since the Bretton Woods Accord put foreign exchange on hold between 1944 and 1971. But we can figure cycles on shorter periods.

Let's look at a cycle in more detail. Figure 1.8 shows a series of three cycles. The top of the cycle (C) is called the *crest* and the bottom (T) is known as *trough*. Analysts measure cycles from trough to trough.

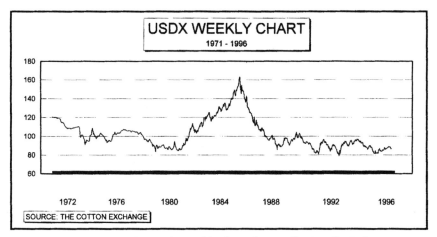

Figure 1.7. Long-term cycles on the weekly chart of FINEX® US Dollar Index. (*Source:* The Cotton Exchange.)

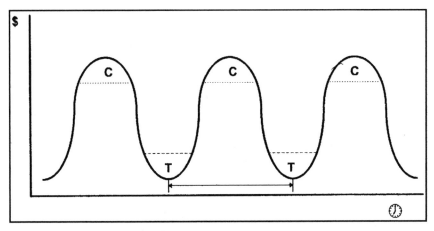

Figure 1.8. The structure of cycles.

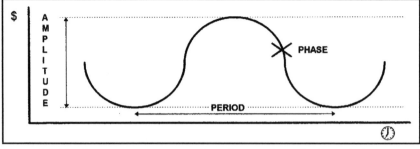

Figure 1.9. The three gauging measures of a cycle: period, amplitude, and phase.

Cycles are gauged in terms of amplitude, period, and phase. As illustrated in Figure 1.9:

- The *amplitude* shows the height of the cycle.
- The *period* shows the length of the cycle.
- The *phase* shows the location of a wave trough.

As shown in the weekly chart of FINEX® US Dollar Index, cycles are not symmetrical, and the bull and bear markets generate different patterns. Bull markets have the crests of their cycles biased to the right, and this pattern is called *right translation.* Conversely, bear markets have the troughs of their cycles leaning to the left—*left translations*—and they last longer. In Figure 1.10, the cycle analysis is displayed at the bottom of the chart.

Types of Charts: A Comparison

The information condensed in technical analysis charts enables the skillful chartist to draw profitable signals for future price activity. Chartists use four types of charts:

- *Line chart* (see Figure 1.11).
- *Bar chart* (see Figure 1.12).
- *Candlestick chart* (see Figure 1.13).
- *Point and figure chart* (see Figure 1.14).

Line Chart

The *line chart* is the original type of chart. To plot it, single prices for a selected time period are connected by a line. The most popular line chart is the *daily chart.* Although any point of the day can be plotted, most traders focus on the closing price, which they perceive as the most important (see Figure 1.15).

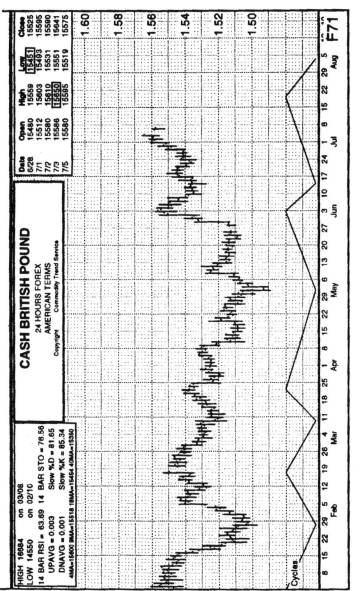

Figure 1.10. Examples of right and left translations. (*Source:* Commodity Trend Service.)

15

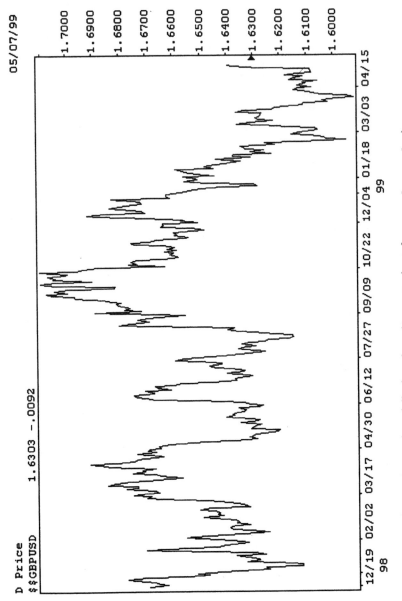

Figure 1.11. The British pound/US dollar line chart. (*Source:* Bridge Information Systems, Inc.)

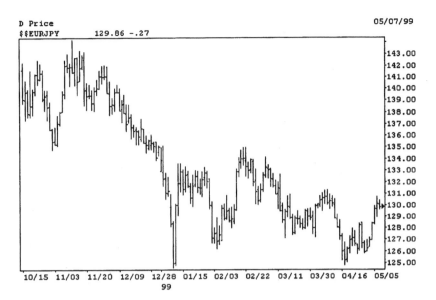

Figure 1.12. The euro/Japanese yen bar chart.
(*Source:* Bridge Information Systems, Inc.)

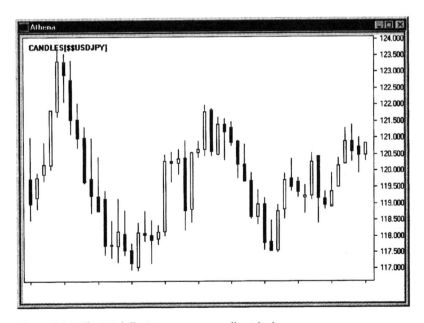

Figure 1.13. The US dollar/Japanese yen candlestick chart.
(*Source:* Bridge Information Systems, Inc.)

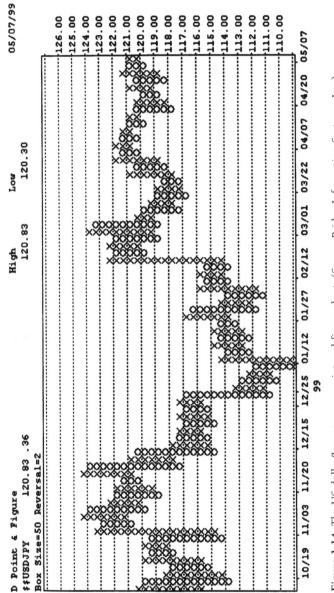

Figure 1.14. The US dollar/Japanese yen point and figure chart. (*Source:* Bridge Information Systems, Inc.)

18

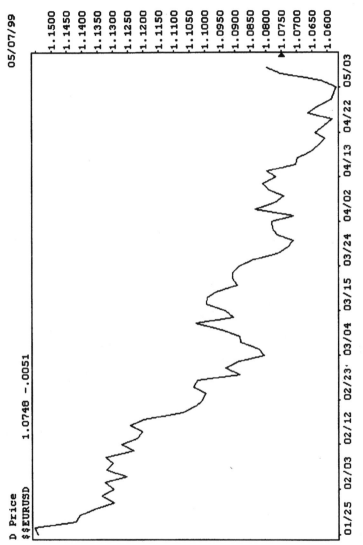

Figure 1.15. The euro/US dollar line chart. (*Source:* Bridge Information Systems, Inc.)

An immediate problem with the daily line chart is that it does not show price activity during the day. With so much information missing, should line charts even be considered for technical analysis?

They should be for several reasons. With the sophistication of current charting services, you can capture daily price activity. You can change the time span for which you need to see the price fluctuation to a very short period, such as in a tick chart, and virtually all prices will be plotted for you to analyze (see Figure 1.16). Daily line charts are particularly useful when looking for the big picture—the major trend—because a bar chart showing intraday activity may generate false signals. Also, when plotted over a long stretch of time, like several years, a line chart is easier to read than other charts (see Figure 1.4), and it is better suited for use as the building block for other studies.

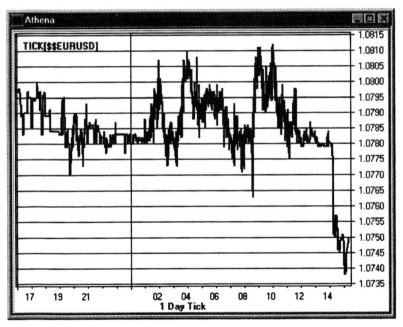

Figure 1.16. The euro/US dollar tick chart.
(*Source:* Bridge Information Systems, Inc.)

However, just like the point and figure chart, the line chart is a continuous chart. This is a disadvantage because price gaps cannot be charted on continuous charts. (Gaps are discussed in Chapter 4.)

Bar Chart

The *bar chart* is arguably the most popular type of chart currently in use. It consists of four significant points:

- The *high* and the *low prices,* which are united by a vertical bar.
- The *opening price,* which is marked with a little horizontal line to the left of the bar.
- The *closing price,* which is marked with a little horizontal line to the right of the bar (see Figure 1.17). The opening price is not always important for analysis.

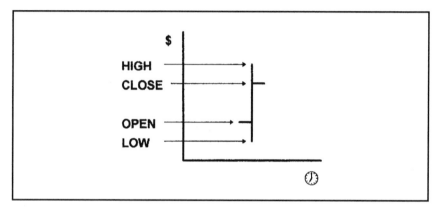

Figure 1.17. The price structure of a bar.

Bar charts have several advantages. An obvious one is that they display the currency range for the selected time period. While any period may be selected, the most popular period is daily, followed by weekly.

Another advantage is that, unlike the line and point and figure charts, the bar chart is able to reflect price gaps, which are formed in the currency futures market. Although the currency futures market is trading around the clock, the currency futures market is physically open only for about a third of the trading day (Chicago International Monetary Market is open for business 7:20 AM to 2:00 PM CDT). Therefore, price gaps may occur between two days' price ranges. Incidentally, the bar chart is the chart of choice among currency futures traders.

Bar charts, however, are unable to plot the entire price fluctuation, even when plotted for short time periods (see Figure 1.18).

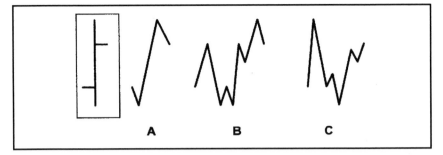

Figure 1.18. Examples of possible market fluctuations hidden by a bar chart.

Candlestick Chart

The *candlestick chart* was developed in Japan, probably around 1750. Despite its venerable age, it was only in the 1980s that it became popular among non-Asian traders. This exposure was by and large possible because of the breakthroughs in electronic charting. Moreover, Steve Nison's book, *Japanese Candlestick Charting Techniques,* enabled more people to understand this type of chart.

The candlestick chart is closely related to the bar chart. The structure of the candlestick also consists of four major prices: high, low, open, and close (see Figure 1.19). At the most basic level, candlestick charts are easier to view than bar charts. In addition, candlestick charts provide particular trading signals.

The *body,* or *jittai,* of the candlestick bar is formed by the opening and closing prices. To indicate that the opening was lower than the closing, the body of the bar is left blank, as you can see in Figure 1.19A. In its original form, the body was colored red. However, current standard electronic displays allow you to keep it blank or select a color. When the currency closes below its opening, the body is filled, as in Figure 1.19B. In its original form, the body was colored black. Again, today's electronic displays allow you to keep it filled or select a color of your choice.

A candlestick chart therefore clearly shows the intraday (or weekly) direction. When the high and the low differ from the opening and closing levels, the rest of the range is marked by two "shadows": the *upper shadow,* or *uwakage,* and the *lower shadow,* or *shitakage.* For illustration purposes, assume the price activity shown in Figure 1.20. In example A, the GBP/USD opened at 1.7000 and closed at 1.7200. The high was 1.7230 and the low was 1.6980. In example B, the GBP/USD opened at 1.7200 and closed at 1.7000. The high was 1.7230 and the low was 1.6980.

Just as with the bar chart, the candlestick chart is unable to trace every price movement during a day's activity. We will discuss this type of chart in Chapter 6.

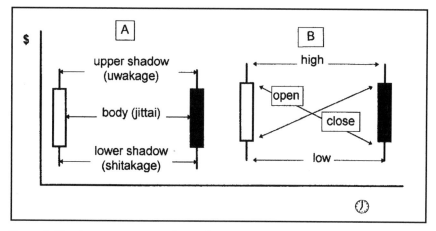

Figure 1.19. The price structure of a candlestick.

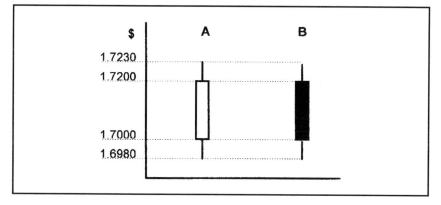

Figure 1.20. Practical examples of a candlestick's price structure.

Point and Figure Chart

The *point and figure chart* takes a different approach from all other types of charts, which have one thing in common: Prices are always plotted against certain periods of time. The point and figure chart completely disregards time, concentrating fully on the price activity. When the currency moves up, the fluctuations are marked with Xs. Moves on the downside are plotted with Os. This chart was also designed to minimize the amount of statistical noise. The direction on the chart changes only if the currency is reversed by a certain number of pips. Prices are divided in boxes, and each box contains an equal number of pips. In foreign exchange, it is common to allocate 10 pips per box.

A very popular point and figure chart is 1 × 3 (see Figure 1.21), which means that as long as the currency continues in the same direction,

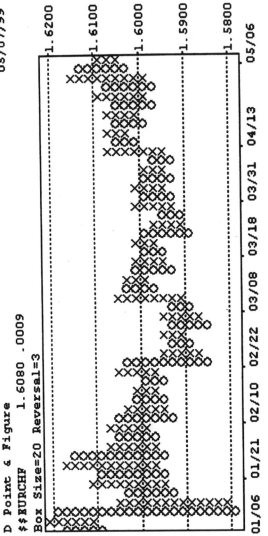

Figure 1.21. The euro/Swiss franc point and figure chart. (*Source:* Bridge Information Systems, Inc.)

every single box is recorded. The currency must reverse by three boxes before they are plotted. In this type of chart, minor activities are ignored, allowing the trader to concentrate solely on price fluctuations.

This chart is very popular among the intraday futures traders because the trading signals are easier to see and more precise than on other charts; no personal interpretation is necessary. It has become more popular since the late 1980s, due to the electronic charting exposure.

Point and figure charting will be discussed in detail in Chapter 5.

In conclusion, the different types of charts require different amounts of information for plotting. The line chart requires only one price per period, while the bar and candlestick charts need four prices per period, and the point and figure chart requires all prices but no time frame. The information you need to plot the four types of charts is summarized in Figure 1.22. If you want to compare all these types of charts, you can plot them together on the same screen. Figure 1.23 shows you an example of the euro/Japanese yen plotted four different ways.

Type of Chart	Price	Period
Line	1	1
Bar	4	1
Candlestick	4	1
Point and figure	All	0

Figure 1.22. The information required to plot each type of chart.

Arithmetic and Logarithmic Scale

Most charts are plotted on an arithmetic scale—"as is," if you wish. Some chartists prefer their charts plotted on a logarithmic scale, to emphasize the percentage price change. This feature may come in handy either for long-term charts or for currencies that have staged significant price changes. Figures 1.24 and 1.25 show you line charts of the same currency, the euro/Japanese yen, plotted over the same period first on an arithmetic scale and then on a logarithmic scale. Which do you prefer?

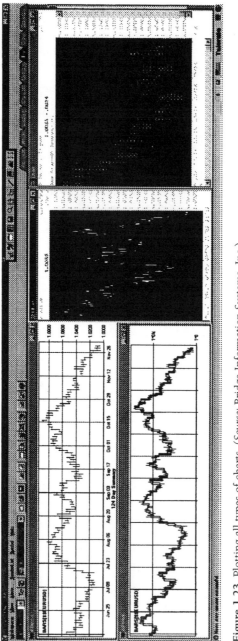

Figure 1.23. Plotting all types of charts. (*Source:* Bridge Information Systems, Inc.)

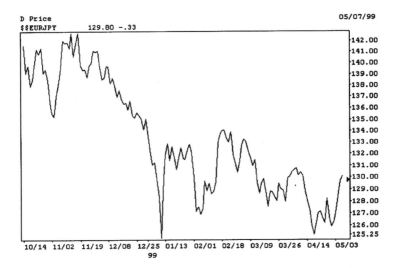

Figure 1.24. A line chart of the euro/Japanese yen plotted on a logarithmic scale. (*Source:* Bridge Information Systems, Inc.)

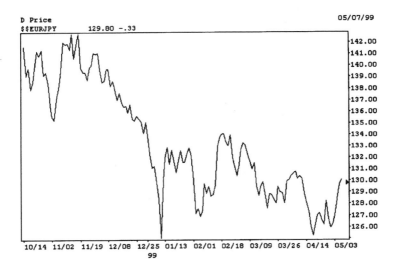

Figure 1.25. A line chart of the euro/Japanese yen plotted on an arithmetic scale. (*Source:* Bridge Information Systems, Inc.)

Volume and Open Interest

Volume consists of the total amount of currency traded within a period of time, usually one day. For example, by 1999, the total foreign currency daily trading volume exceeded $1.4 trillion (see Figure 1.26). Traders are naturally less interested in the overall volume than in the volume for a specific currency. A large trading volume suggests that there is high interest and liquidity in a market. In addition, certain chart formations require heavy trading volume for successful development. An example is the head and shoulder formation, a pattern presented in Chapter 2.

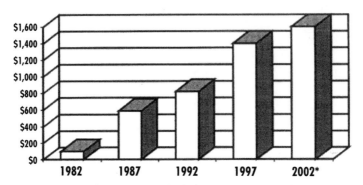

Figure 1.26. Daily turnover in the foreign exchange market in billions of dollars. *Estimated. (*Source:* Bank For International Settlements)

A low volume raises a red flag and warns the trader to veer away from the market. Charts that show seemingly large price jumps may tempt traders. When this price action occurs on low volume, however, you should never get involved. In a low-volume market, you can easily get an entry price but you are pressed hard when you want to exit. There is a high probability that you will get a rather unfavorable rate.

Despite its natural importance, volume is not easy to quantify in most of the foreign exchange markets. Volume figures can be calculated for the futures and for the options on futures because both take place on centralized exchange floors and all the trades go through the clearinghouse. However, these instruments barely represent about 3% of the currency markets. In the spot, forwards and cash options markets, the trading is completely decentralized, and volume is therefore impossible to gauge.

To minimize this problem, traders have to learn how to estimate volume. One method is to extrapolate the figures from the futures market. The downside to this approach is that you generally have to wait until tomorrow to get today's figures. Another method is to "feel" the size of volume based on the number of calls on the dealing systems or phones, as well as on the "noise" from the brokers' market. It isn't the easiest job in the world, but who ever thought it would be?

Open interest is the total exposure, or net outstanding position, in a certain instrument. Generally, the open interest represents the difference between the outstanding long positions and the outstanding short positions. Should you need this information, you may want to look at both figures on an individual basis.

Open interest is plagued by the same problems as volume. As already mentioned, the figures for volume and open interest are available solely for currency futures and options on currency futures.

If you have access to printed or electronic charts on futures, you will be able to see these numbers plotted at the bottom of the futures charts. In Figure 1.27, volume is represented by a bar chart, and the open interest is plotted as a line chart.

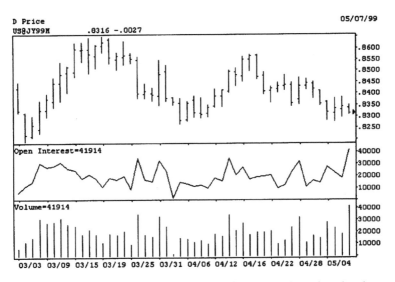

Figure 1.27. A line chart of the Japanese yen futures bar chart plotted with volume and open interest. (*Source:* Bridge Information Systems, Inc.)

How does the open interest data affect your currency forecast? Figure 1.28 shows you the correlations between currency prices and open interest figures, along with the resulting trading signals.

Prices	Open Interest	Signal
Up	Up	Up
Flat after up	Flat after up	Bearish reversal
Down	Up	Down
Up	Too high	Bearish reversal
Flat	Up	Potential breakout
Up	Down	Down
Down	Up	Down
Down	Down	Down until reaching intrinsic value

Figure 1.28. The correlations between currency prices and open interest.

Open interest figures from the futures market are not foolproof. The foreign exchange markets are a *zero-sum game* (particularly in the futures markets, where every outstanding contract must be offset by an opposite contract). So, on an individual basis, the figures may be misleading. For instance, suppose an investment bank or a fund opens a position in the transparent futures market by buying a large amount of JPY futures. Is that player long JPY? The position may also be a hedge for an option position or a cash position, or simply a strategy of throwing the dogs off the scent by showing one hand where everybody sees it, and doubling it up on the opposite side in the cash market.

Trend

Understanding the idea of trend is essential to technical analysis. "The trend is my friend" is a familiar quote that is deeply rooted in the market experience, and it should be respected and attentively observed. A trend simply shows a major direction of the market. Therefore, a *trend* may be:

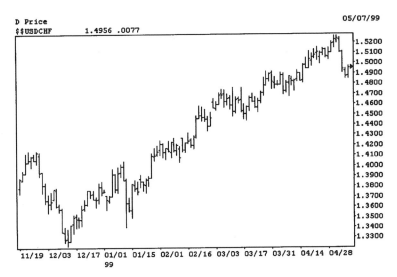

Figure 1.29. An up trend in the US dollar/Swiss franc.
(*Source:* Bridge Information Systems, Inc.)

1. *Upward:* The currency makes consecutive higher highs and higher lows. In Figure 1.29 you see an example of the long-term up trend in the US dollar/Swiss franc.

2. *Downward:* The currency registers lower highs and lower lows. Figure 1.30 illustrates a down trend in the US dollar/Japanese yen.

3. *Sideways,* also known as a "flat market " or "trendless": The currency is locked in a tight horizontal range. Figure 1.31 shows tight sideways trading range in the US dollar/Japanese yen between August 1994 and February 1995.

Since the markets do not move in a straight line in any direction, but rather in zigzags, the direction of these peaks and troughs create the market trend.

In addition to direction, trends are also classified by time frame:

- *Major* or *long-term trends.*
- *Secondary* or *medium-term trends.*
- *Near-term* or *short-term trends.*

Any number of secondary and near-term trends may occur within a major trend.

Figure 1.30. A down trend in the US dollar/Japanese yen.
(*Source:* Bridge Information Systems, Inc.)

The time frames for the types of trends vary widely. Although Charles Dow suggested that a major trend lasts one year, currently the time span seems to be between one year and seven years. Secondary trends should be measured in terms of months, and minor trends continue for only a matter of weeks.

The significance of trends is a function of *time* and *volume*. The longer the prices bounce off the support and resistance levels (to be discussed later in this chapter), the more significant the trend becomes. Trading volume is also very important, especially at the critical support and resistance levels. When the currency bounces off these levels on heavy volume, the significance of the trend increases.

Trend Lines

A *trend line,* the natural result of tracking a trend, consists simply of a straight line connecting the significant highs (peaks) or the significant

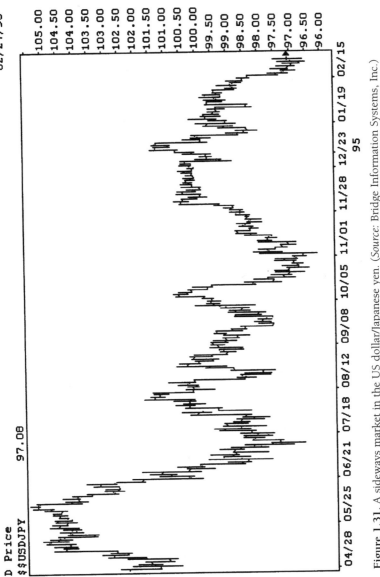

Figure 1.31. A sideways market in the US dollar/Japanese yen. (*Source:* Bridge Information Systems, Inc.)

lows (troughs). Following the possible trend directions, the trend lines may be classified as:

1. *Rising trend lines* (see Figure 1.32).
2. *Declining trend lines* (see Figure 1.33).
3. *Flat trend lines* (see Figure 1.34).

Figure 1.32. A rising trend line in the US dollar/Japanese yen.
(*Source:* Bridge Information Systems, Inc.)

Drawing the trend line is easy, since only two points are necessary. But this line is tentative, and you must confirm it with a third contact point. Once the trend seems to be securely set in its tracks, you must recall an important point made in the beginning of this chapter: *Financial markets are trend followers*. Therefore, we can now expect the currency to *maintain the general direction* and *velocity*.

The most significant trend lines occur around an angle of 45 degrees (see Figure 1.35). This important technique was established by W. D. Gann, the renowned investor and technical analyst. He noted that a trend

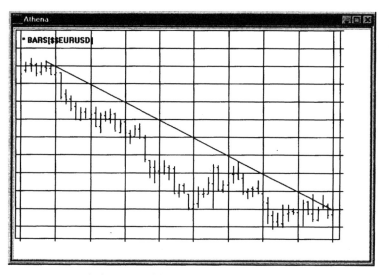

Figure 1.33. A declining trend line in the euro/US dollar.
(*Source:* Bridge Information Systems, Inc.)

line at a sharper angle suggests that the rally is unsustainable. Conversely, a trend line at less of an angle indicates that the trend is close to reverse. He also noted that a longevity of one month or more provides the trend line with increased weight.

We would like to identify not only as many significant trend lines as possible, but also clear support and resistance levels. However, since the market has a life of its own, this is not possible. Many times we see minor trend line penetrations, but, as a rule, we should disregard such minor breakouts.

Trading Signals for Trend Lines

1. Anemic, that is, low-volume breakouts or penetrations, must be disregarded. They occur only because a handful of players either were forced to cut their losses or attempted to spearhead a new trend but failed to stir up sufficient market interest. As you can see in Figure 1.36, these meek breakouts fail quickly and the trend continues unabated.

2. A breakout from an upward trend line must be confirmed by one close below the trend line. Conversely, a breakout from a down

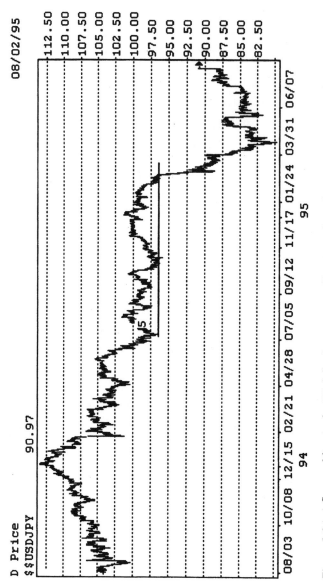

Figure 1.34. A flat trend line in the US dollar/Japanese yen. (*Source:* Bridge Information Systems, Inc.)

36

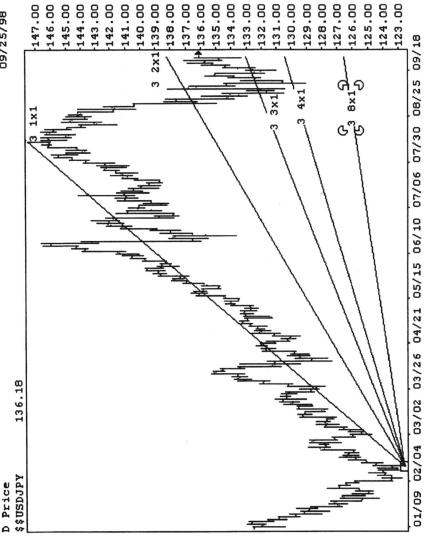

Figure 1.35. The Gann fan lines marked in the US dollar/Japanese yen.
(*Source:* Bridge Information Systems, Inc.)

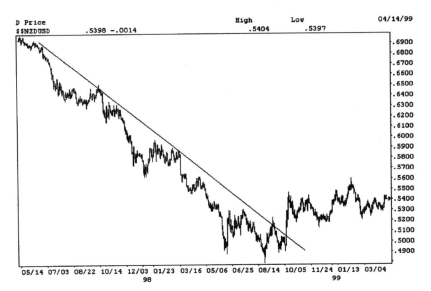

Figure 1.36. Weak violations of the trend line in the New Zealand dollar/US dollar. (*Source:* Bridge Information Systems, Inc.)

    trend line should be confirmed by one close above the trend line (see Figure 1.37).

3. In a less decided market, a breakout may have to be followed by two consecutive closes outside the trend line before the penetration is confirmed (see Figure 1.38).

4. Use a price filter of 1% to test the validity of the breakout.

5. After confirmation, breakouts are likely to be followed by a period of consolidation (see Figure 1.39). It is relatively rare for trends to suddenly reverse direction.

6. The longer the consolidation periods last, the steeper the following rallies will be (see Figure 1.40), because consolidations are rarely profitable and there is a subsequent rush to make money.

7. Breakouts through upward trend lines tend to test the strength of the former support line, now turned into a resistance line. Conversely, breakouts through downward trend lines tend to test the former resistance line, now turned into a support line (see Figure 1.41).

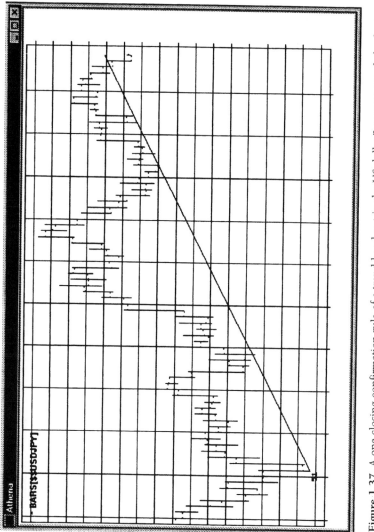

Figure 1.37. A one closing confirmation rule of a trend breakout in the US dollar/Japanese yen daily chart. (*Source:* Bridge Information Systems, Inc.)

Figure 1.38. Two or more closings confirmation rule of a trend breakout in the euro/Canadian dollar daily chart.
(*Source:* Bridge Information Systems, Inc.)

Figure 1.39. Consolidation outside the trend line following a trend breakout in the Australian dollar/US dollar daily chart.
(*Source:* Bridge Information Systems, Inc.)

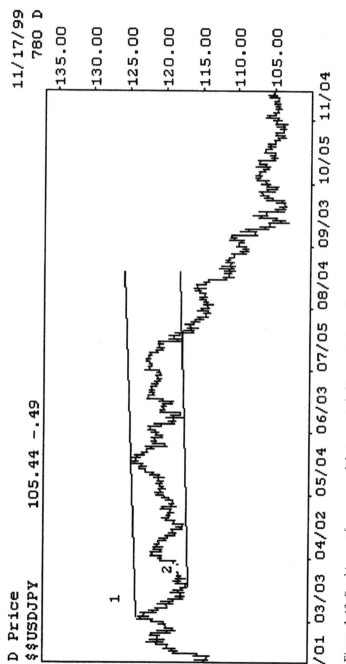

Figure 1.40. Breaking out from a consolidation period. (*Source:* Bridge Information Systems, Inc.)

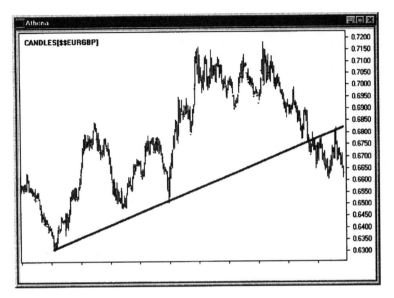

Figure 1.41. Currency testing the strength of the new resistance line following a breakout of a rising trend line. (*Source:* Bridge Information Systems, Inc.)

The Channel Line

A *channel line* is a parallel line that you can trace against the trend line, connecting the significant peaks in an up trend and the significant troughs in a down trend (see Figure 1.42). Along with the trend line, the

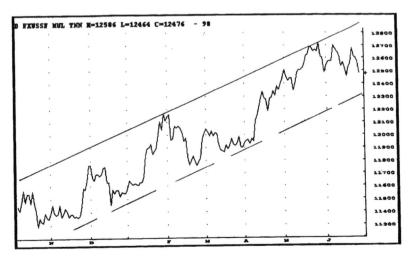

Figure 1.42. Channel lines, which are marked with dashed lines. (*Source:* FutureSource copyright © 1986-1996. All rights reserved.)

channel line creates a channel that borders the currency trend. In a down trend, the channel line is at the bottom of the channel (the support line). In an up trend, the channel line is at the top of the channel (the resistance line). Channel lines do not have to be parallel lines to the trend lines. In fact, they may not occur at all because the market conditions and the degrees of price volatility vary.

Trading Signals for Channel Lines

1. A channel is an attractive chart pattern for traders, since the number of buying signals is approximately doubled by the number of selling signals. The currency price basically vacillates between the trend line and channel line.

2. The failure of price to reach the trend line should be interpreted as a possible trend acceleration (see Figure 1.43).

Figure 1.43. The currency's failure to reach its rising trend line was a signal that the up trend is accelerating. (*Source:* Bridge Information Systems, Inc.)

3. The failure of price to reach the channel line should be construed as a case of a weakening trend (see Figure 1.44).

4. The break of the channel line confirms a trend acceleration (see Figure 1.45).

Figure 1.44. The currency's failure to reach its declining channel line was a signal that the down trend is weakening.
(*Source:* Bridge Information Systems, Inc.)

Figure 1.45. The breakout of the US dollar/Canadian dollar above the rising channel line confirmed the acceleration of the trend.
(*Source:* Bridge Information Systems, Inc.)

5. A channel breakout suggests a currency target price equal to the width of the channel.

6. Generally, consolidation does not follow the breakout of the channel line. (This is an optimistic breakout.) In a bullish breakout, the

long positions increase and the neutral positions go long. Conversely, the bearish breakout of the channel line triggers an increase in the short positions and convinces neutral traders to sell the currency short.

Support and Resistance Levels

The peaks representing the price levels at which the selling pressure exceeds the buying pressure are known as *resistance levels*. The troughs representing the levels at which the selling pressure succumbs to the buying pressure are called *support levels*. You can see several support and resistance lines in Figure 1.46. In an up trend, the consecutive support and resistance levels must exceed each other, respectively. The reverse is true in a down trend. Although minor exceptions are acceptable, these failures should be considered as warning signals for trend changing.

The importance of the support and resistance levels goes beyond their original functions. If these levels are convincingly penetrated, they tend to turn into their opposites. A firm support level, once it is penetrated on heavy volume, is likely to turn into a strong resistance level (see Figure 1.47). Conversely, a strong resistance turns into a firm support after being soundly penetrated (see Figure 1.48).

Speedlines

An innovative approach to trend analysis, based on the range of trends, was perfected by Edson Gould and was quickly embraced by the technicians. It is called *speedlines* (see Figure 1.49). To calculate speedlines, divide the total range of a trend into thirds on a vertical line, which originates at the bottom of the range for an up trend and at the top of the trend for a down trend. In a down trend, the first resulting speedline (A) is plotted by using as coordinates the origin and the 1/3 of the range from the bottom. The second speedline (B) connects the origin and the price at the 2/3 level of the range.

In an up trend, price fluctuations away from the trend that find support at the first (1/3) speedline indicate that the trend is going to continue smoothly. The penetration of the first (1/3) speedline is a trend weakening warning, and the next level to watch is the second speedline (2/3). As the first (1/3) speedline becomes resistance, the currency price

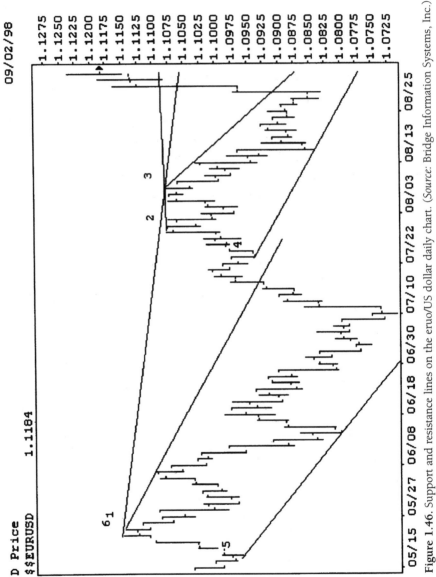

Figure 1.46. Support and resistance lines on the eruo/US dollar daily chart. (*Source:* Bridge Information Systems, Inc.)

Figure 1.47. A strong support line turns into a strong resistance line after breakout. (*Source:* Bridge Information Systems, Inc.)

Figure 1.48. A strong resistance line turns into a strong support line after breakout. (*Source:* Bridge Information Systems, Inc.)

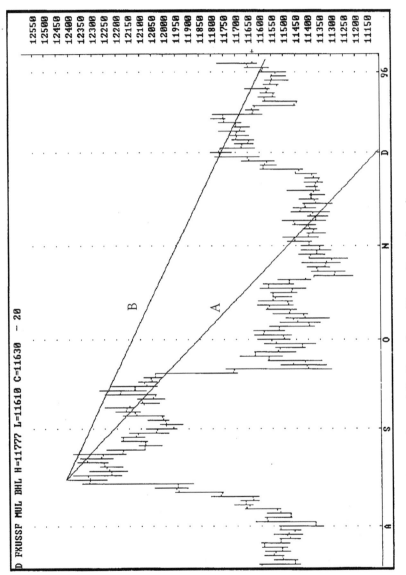

Figure 1.49. Speedlines in the US dollar/Swiss franc. (*Source:* FutureSource.)

tends to consolidate between the first (1/3) and second (2/3) speedlines. If the second (2/3) speedline gives way as well, then the next target is the range origin, and the trend is confirmed to be broken.

Speedlines may be applied to all types of charts and for all time frames. However, they present the most significance in the short- and medium-term bar charts and long-term line charts.

Andrews' Pitchfork

Andrews' Pitchfork is a method of channel identification that was developed by Alan Andrews. To plot it, use a diagonal line to connect a significant high (point A in Figure 1.50) and a significant low (point B) prior to an emerging trend. Then you bisect the diagonal line at point C. You create the pitchfork by inserting a trend line, a channel line, and a median line. In effect, you are splitting the major channel into two minor equidistant channels. The median line should provide additional support/resistance levels. If you use an electronic charting service, select the study, click on a high point (A), a low point (B), and on the beginning of the trend (C). Andrew's Pitchfork will appear on your chart in a jiffy.

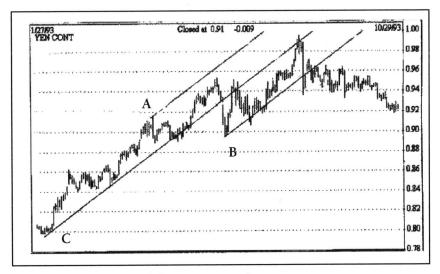

Figure 1.50. Andrews' Pitchfork in Japanese yen futures.
(*Source:* Stratagem Software International.)

The Importance of the Long-Term Charts

Although, in terms of time frame, the daily charts are the most popular, the longer-period charts—that is, *weekly* and *monthly charts*—are the most important.

Why are the weekly and monthly charts so useful?

1. These charts make it possible to *compress very long-term information into a single chart* (Figure 1.51).

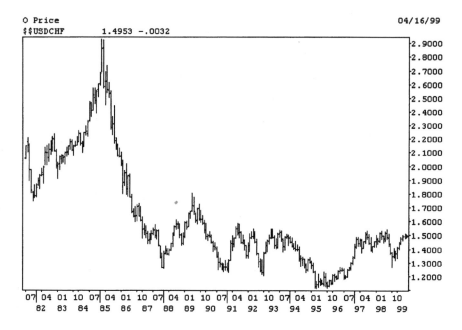

Figure 1.51. The monthly chart of the US dollar/Swiss franc.
(*Source:* Bridge Information Systems, Inc.)

2. Very long-term charts provide extremely important information regarding the *long-term trends or cycles*. Traders can get a correct perspective on the real direction of the market in the long run, the strength or direction of the current trend within that trend (Figure 1.52), or the possibility of breakout from the long-term trend. For

instance, in Figure 1.53 you can see a long-term triple top (a trend reversal formation) in the British pound, which was approaching completion in September 1992. The massive sell-off triggered by George Soros helped the British pound break the support line. This pattern was not evident on short-term charts. We will analyze this type of formation in Chapter 2.

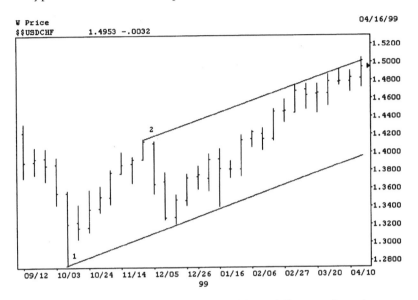

Figure 1.52. An easy-to-spot trend on the weekly US dollar/Swiss franc chart. (*Source:* Bridge Information Systems, Inc.)

Percentage Retracements

As you are well aware of by now, foreign currencies, like all other financial instruments, cannot move straight up or down even in the healthiest of the trends. The market withdraws on profit taking, attracting further buyers. These retracements have been noticed statistically and named percentage retracements. Traders watch several percentage retracements, in search of price objectives. The typical percentage retracements are:

1. *Traditional (Charles Dow):* The *traditional retracement percentages* were developed by Charles Dow at the turn of the last century. They are 1/3, 1/2, and 2/3, or *33%, 50%, and 66%.* A retracement past 66% is considered to be a trend failure (see Figure 1.54).

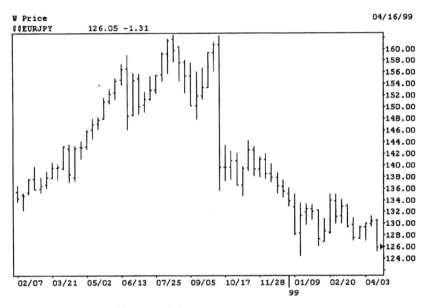

Figure 1.53. The weekly chart of the euro/Japanese yen.
(*Source:* Bridge Information Systems, Inc.)

2. *The Gann percentages:* The *Gann percentages* attach importance to the 1/8 breakdowns. The Gann theory focuses mostly on the 3/8, 4/8, and 5/8, or *38%, 50%, and 62%* retracement figures (see Figure 1.55). Gann's approach to the market is discussed in detail in Chapter 10.

3. *The Fibonacci ratios:* The *Fibonacci ratios* are very popular among Elliott Wave students. These ratios are .382 and .618, or approximately *38% and 62%* (see Figure 1.56).

    Another way using these ratios is the *Fibonacci fans.* As in the construction of speedlines, you measure the range between a significant low and a significant high. Then you deduct .382 and .618 of the range. Finally, you connect the significant low in an up trend, or the significant high in a down trend, with the Fibonacci retracement points. Thus you divide this range into three segments, where the fan lines act as strong support and resistance lines. You may repeat the process with a line to .50 retracement level. The Fibonacci fans are illustrated in Figure 1.57.

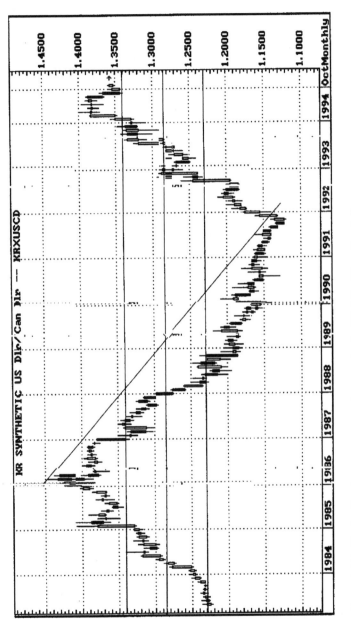

Figure 1.54. Charles Dow's retracement percentages in the US dollar/Canadian dollar. (*Source:* Bridge Information Systems, Inc.)

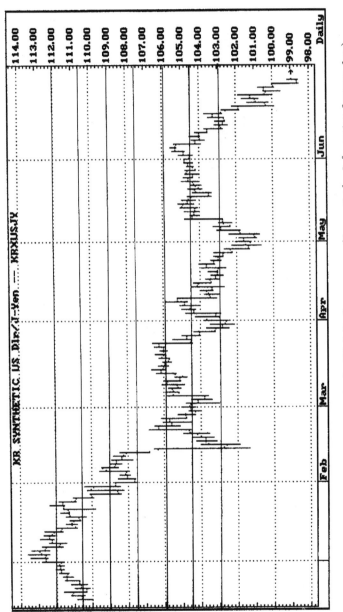

Figure 1.55. Gann percentage retracements in the US dollar/Japanese yen. (*Source:* Bridge Information Systems, Inc.)

Figure 1.56. Fibonacci retracement levels on the US dollar/Canadian dollar. (*Source:* Bridge Information Systems, Inc.)

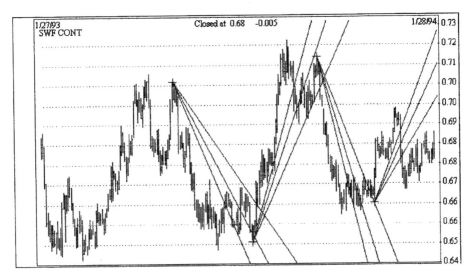

Figure 1.57. The Fibonacci fans in the IMM Swiss franc futures. (*Source:* Stratagem Software International.)

Finally, you can use the *Fibonacci arcs.* In this study, the length of a trend is divided by the Fibonacci ratios at .382, .50, and .618. With the help of a compass or friendly software, you extrapolate these levels into the future to achieve time targets for retracements. As you can see in Figure 1.58, the market has a memory and the IMM Japanese yen futures rebounded on or within the proximity of the Fibonacci arcs.

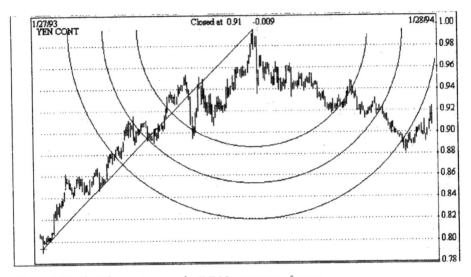

Figure 1.58. The Fibonacci arcs in the IMM Japanese yen futures. (*Source:* Stratagem Software International.)

Fibonacci ratios are discussed in more detail in Chapter 11.

TREND REVERSAL PATTERNS

The ability to identify a trend and properly monitor it is money in the bank. Despite the trend's monetary significance to us (and thus our wish for its eternal life), trends, like many other good things, must come to an end.

This chapter deals with chart patterns that signal trend reversals in the foreign exchange market. Remember from Chapter 1 the "psychological" reaction of traders after the market breaks out of the trend: Most times, they attempt to bring the currency back into the trend, only to be stopped by the old trend line. You will see plenty of proof of this characteristic in most of the reversal patterns. Understanding these formations not only eases the pain of losing your beloved trend, but it also brings you monetary reward as well. After all, if you can't fight them, join them.

Price Structure

The most significant trend reversal patterns are:

1. Head and shoulders and inverse head and shoulders formations.
2. Double top and double bottom formations.
3. Triple top and triple bottom formations.
4. V-formations.
5. Rounded top and rounded bottom formations.
6. Diamond formation.
7. Rectangle formation.

57

The first four patterns—head and shoulders, double tops and bottoms, triple tops and bottoms, and V-formations—are the most significant for technical analysts. Therefore, we call them *major trend reversal formations.*

Head and Shoulders and Inverted Head and Shoulders

Head and Shoulders

One of the most reliable and well-known chart formations, the *head and shoulders* pattern hardly needs an introduction. This pattern consists of three consecutive rallies that occur at the end of an up trend, where the first and third rallies (the *shoulders*) have about the same height and the middle one (the *head*) is the highest. All three rallies are based on the same support, known as the *neckline* (see Figure 2.1).

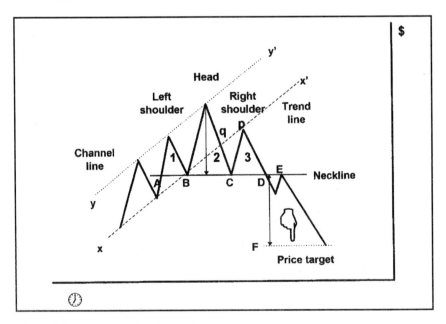

Figure 2.1. A typical head and shoulders pattern.

As you can see in Figure 2.1, the underlying currency, which moved in an up trend, broke the trend line of the *channel xx'* at point *q.* In a typical reaction, the currency rallied back to the previous trend/support line, in an attempt to trade above it. This line turned into a sturdy resistance line and stopped the rally at point *p*. As the currency fell, the trend breakout was confirmed. The last rally, defined by points *CpD* (the *right shoulder*) does not have to occur outside the trend line.

Prior to point A, the neckline was a resistance line. Once the resistance line was broken, it turned into a significant support line. The price bounced off it twice, at points B and C. The neckline was eventually broken at point D under heavy volume, and the trend reversal was confirmed. Once a significant support line is broken, you can expect a retracement to retest the neckline (E), which becomes a resistance line. Since in this case the resistance line held, the price is expected eventually to decline to around level F, which is the *price target* of the head and shoulders formation. The target is approximately equal in amplitude to the distance between the top of the head and the neckline, and it is measured from the point at which the neckline is broken (D).

While there is no strict rule for how long it will take to reach the target, common sense would link it to the duration of development. In any case, the target is unlikely to be reached in a matter of hours.

It is important to *measure the target from the point at which the neckline is broken*. Aspiring technicians tend to measure the target price not only from under the neckline but also from the middle of the formation. This can happen as they measure the height of the head.

A market example of a head and shoulders reversal formation is presented in Figure 2.2. Most head and shoulders formations, of course, look

Figure 2.2. A head and shoulders formation on the euro/US dollar chart. (*Source:* Bridge Information Systems, Inc.)

different from Figure 2.1. Prices fluctuate enough to eliminate any possibility of a clean-looking chart line. Also, the neckline is seldom a perfectly horizontal line.

Signals Generated by the Head and Shoulders Pattern

The head and shoulders formation provides excellent information:

1. At point *q*, the up trend is broken. This tells you about the health of the *trend.*
2. The typical *consolidation period*—between point *q* and the breakout point *D*—is over.
3. There is a new *resistance line.* The neckline, initially acting as a support line, turns into a significant resistance line. After penetrating the neckline at point *D,* the market may retest it at point *E.*
4. The *price direction* has changed. If the neckline holds the buying pressure at point *E,* then the price direction is the opposite of that of the three significant peaks (the head and the two shoulders).
5. The *price target* (the height of the head measured from the breakout point *D*) is confirmed when the head and shoulders formation breaks through the neckline under heavy trading volume.

Heavy market volume at the breakout is the main requirement for the successful development of this formation. Since gauging volume is possible only in the currency futures market, the trader has to estimate volume in the cash market by listening to the trading "noise" or, less commonly, by extrapolating the currency futures' volume. Light volume is a strong warning of a false breakout and a possible sharp backlash in the currency price.

The time frame for this chart's formation is from several weeks to several months. (Intraday chart formations are not reliable.) The longer the formation time is, the more significance you should attach to this pattern.

Potential Problems

1. The height of the shoulders should be about equal. More important, the shoulders must not be taller than the head, as shown at the right.

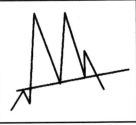

2. It is vital that the significant points *A, B, C,* and *D* are all tangential to the neckline. In the chart to the right, point *C* is not tangential.

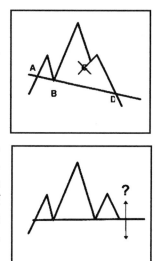

3. The head and shoulders formation is confirmed only when a breach of the neckline follows the completion of the three rallies and their reversals. The failure of the price to *break through the neckline on a closing prices basis* puts the formation on hold or even negates its validity.

Trading Tips

1. The *breakout is confirmed only by a closing price or by several closing prices outside the neckline,* not by daily highs or lows. Price overshooting doesn't buckle trends. Bar and candlestick charts are more difficult to read. Line charts make it easiest for you to identify trend breakouts.

2. *Do not go long for any significant duration on the rise of the right shoulder (Cp in Figure 2.1). Here is your chance to go short (with some tight stops on) before the rest of the market flocks to sell.*

3. *If you take a position before the breakout, don't go short just before the neckline is tested.* If the market penetrates the neckline, you have a big profit. If the market fails, however, your tight stop-loss orders may be impossible to execute due to high trading speed and low liquidity. The risk-reward ratio is not in your favor.

4. *If you take a position after the breakout, sell extremely fast and take your profit within 50 pips from the breakout level of a medium-term pattern.* Price quotes are likely to be very wide and a second's hesitation may mean you sold at the low, since it is common for the market to retest the neckline.

5. *If you take a position before the breakout, sell again on the breakout and quickly place a profit-taking order for about half of your position far from the neckline, about 50 pips or more.* You have an excellent chance to be filled in the shock of the moment, especially on a matching system.

6. Because the market expects imminent completion, there are likely to be large positions outstanding in the direction of the future target. Therefore, *a chart failure is likely to trigger a steep position reversal.*

7. *The target should be considered as simply a suggested objective.* The price only accidentally stops exactly at the target level. Generally, it may fall a little short or overshoot. Fine-tune your profit-taking approach so that you participate in the significant part of the move toward the target. Don't bog yourself down trying to clean out the last few pips.

8. If you are a cash trader, remember both the *importance of volume and the lack of volume information.* Correctly estimating volume in the cash market is a matter of personal skill in a very noisy and confusing environment.

The Inverted Head and Shoulders

The *inverted head and shoulders* formation is a mirror image of the previous pattern. As you can see in Figure 2.3, the underlying currency broke out of the down trend ranged by the *xx'–yy'* channel, at point *j,* after mak-

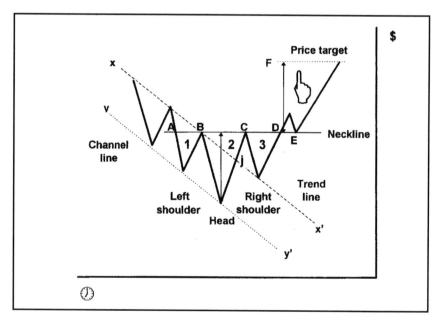

Figure 2.3. A typical inverted head and shoulders.

ing its lowest low. It retested the previous trend line, now turned into a support line (rally number 3). Of the three consecutive rallies, the shoulders (1 and 3) have approximately the same height, and the head is the lowest. Prior to point *A,* the neckline was a support line. Once this line was broken, it turned into a significant resistance line. The price bounced off the neckline twice, at points *B* and *C.* The neckline was eventually broken at point *D,* on heavy volume. Since the significant resistance line was broken, a retracement can be expected to retest the neckline (*E*), now a support line again. If it holds, the price is expected eventually to rise to around level *F,* which is the price target of the inverted head and shoulders formation.

The price objective is approximately equal in amplitude to the distance between the top of the head and the neckline, and is measured from the breakout point *D.*

Figure 2.4 presents an example of an inverse head and shoulders pattern at work in the British pound/US dollar market.

Double Top and Double Bottom

Double Top

Another very reliable and common trend reversal chart formation is the *double top pattern.* As its name clearly and succinctly describes, this pattern consists of two tops (peaks) of approximately equal heights. As you can see in Figure 2.5, a currency in an up trend broke the trend line *xx'* at point *s,* just after reaching its highest peak (*B*). The market tried to extend the up trend, but the currency failed to make a new high. In fact, this last rally was stopped at point *D* by the previous trend line, now turned into a strong resistance line.

A line is drawn below and parallel to the resistance line that connects the two tops. The characteristics of this line are identical to the head and shoulders' neckline. The neckline is a strong support for price level *C* but eventually fails at point *E.* The support line turns into a resistance line, which holds the market backlash at point *F.* The price objective is at level *G,* which is the average height of the double top formation, measured from point *E.* (*Note:* The term "neckline" is generally used for head and shoulders formations. I also use this term for double tops and bottoms, as well as for triple tops and bottoms, because the line serves the same technical function.)

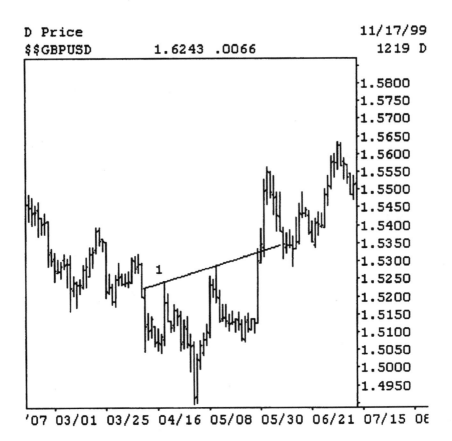

Figure 2.4. Example of an inverted head and shoulders pattern.
(*Source:* Bridge Information Systems, Inc.)

Signals Provided by Double Tops

The double tops formation provides the following information:

1. The *support line* is set between points A and E.
2. The *resistance line* is set between points B and D.

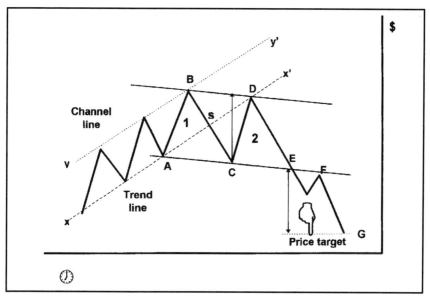

Figure 2.5. A typical double top formation.

3. The second *resistance line* is set between points E and F. This is the previous support line *AE*.
4. If the neckline holds the buying pressure at point F, the *price direction* is opposite that of the peaks (*bearish*).
5. The *price target* is confirmed when the price breaks through the neckline on heavy trading volume.

As in the head and shoulders pattern, heavy market volume on the breakout through the neckline is a vital requirement for the successful completion of this formation. Again, since gauging volume in traditional ways is possible only in the currency futures market, the trader must estimate the market volume based on the trading "noise." Light volume is a strong case for a false breakout, which generally triggers a sharp backlash in the currency price.

The time frame for this chart's formation runs from several weeks to several months. Intraday chart formations are less reliable. There is a strong correlation between the length of time the pattern takes to develop and the significance of the formation.

The time needed to reach the price objective is unclear. Generally, the longer the pattern takes to develop, the longer it should take to reach the price objective.

Measure the target from the point at which the neckline is broken. Avoid the trap of measuring the target price from the middle of the formation under the neckline, which can happen as you measure the average height of the formation.

Most double top formations look different from that in Figure 2.5. As shown in Figure 2.6, prices fluctuate enough to create a "weathered" looking line, and the neckline is seldom perfectly horizontal.

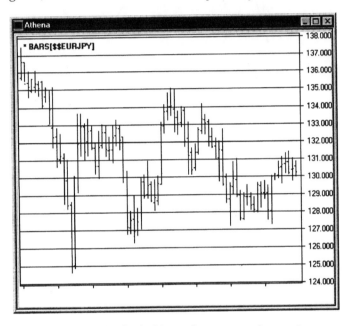

Figure 2.6. Example of a double top formation on the euro/Japanese yen chart. (*Source:* Bridge Information Systems, Inc.)

Potential Problems

1. The height of the peaks should be about equal. The formation to the right is not a double top.

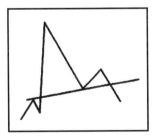

2. The significant points *B* and *C* must be tangential to the neckline. The failure of this requirement (at *C*) nullifies the characteristics of the formation.

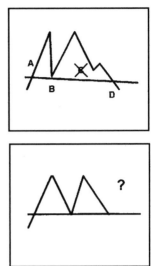

3. The double tops formation is confirmed only when the neckline is breached (i.e., the closing price is outside the neckline) after the two rallies and their respective reversals are completed. The failure of the price to break through the neckline puts the formation on hold or negates its validity.

Trading Tips

1. Remember that the breakout confirmation consists of a *daily closing price outside the neckline.*
2. *Do not go long for any significant period of time on the rise of the second top (CD in Figure 2.5). Here is your chance to go short* (with some tight stops on) before the rest of the market sells.
3. The majority of the market will wait for the confirmation of the neckline's breaching before attempting to jump on the bandwagon. *If you open a new short position after the breakout, be very fast and decisive.* The prices will be widespread and highly volatile, making it difficult for you to join in. Hesitation may make you sell too close to the price objective and probably force you to cut your losses around the neckline since the market quite possibly will retest it.
4. *If you take a position prior to the breakout, sell again on the breakout and quickly place a profit-taking order for half of your position far from the neckline, about 70 pips or more.* You have a good chance to be filled in the shock of the moment, especially on a matching system.
5. Given the expectation of imminent completion, large positions may be outstanding in the direction of the future target; so a chart failure is likely to trigger a steep position reversal. If you are short prior to

the breakout and want to execute your stop-loss orders, remember that they are *nearly impossible to execute* due to the high speed and low liquidity of the market.

6. *Consider the price objective as simply a suggested target.* Generally, the price either falls short or overshoots. Evaluate any additional information to fine-tune the profit-taking level.

7. Remember that *volume is very important but it is the one bit of information that you lack.* Estimating volume correctly in the cash market shows *your personal degree of skill.*

Double Bottom

The *double bottom* formation is a mirror image of the double top pattern: You can apply the same characteristics, potential problems, and signals. As shown in Figure 2.7, the bottoms have about the same amplitude. A parallel line (the neckline) is drawn against the line connecting the two bottoms (*B* and *D*). As a support line, it is broken at point *A*. It turns into a strong resistance for price level *C* but eventually fails at point *E*. The resistance line turns into a strong support line, which holds the market

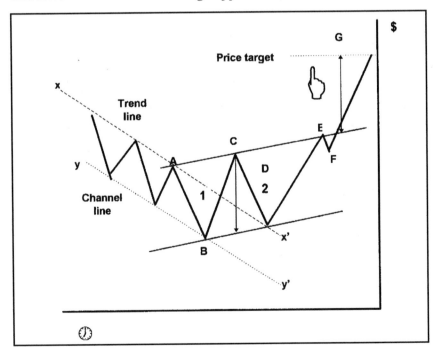

Figure 2.7. A typical double bottom formation.

backlash at point F. The price objective is at level G, which is the average height of the bottoms, measured from point E.

Figure 2.8 presents an example of a double bottom pattern at work in the foreign exchange markets.

Triple Top and Bottom

Triple Top

The *triple top* is a hybrid of the head and shoulders and double top trend reversal formations. As shown in Figure 2.9, in a typical triple top formation, the tops have about the same height. A parallel line (the neckline) is drawn against the resistance line connecting the three tops (B, D, and F). The neckline, acting as a resistance line, is broken at point A. It turns into a strong support for price levels C and E, but eventually fails at point G. The support line turns into a strong resistance line, which holds the market backlash at point H. The price objective is at level I, which is the average height of the three top formation, as measured from point D.

Like the double top, the formation fails at point E. The price moves up steeply toward point F. The resistance line is holding once more and the price drops sharply again toward point G. At this level, the market pressure enables prices to penetrate the support line. After a possible retest of the neckline, prices drop further, eventually to reach the price objective.

Triple Bottom

Conversely, the *triple bottom formation* is a hybrid of the inverted head and shoulders and double bottom formations. Apply the same characteristics, potential problems, signals, and the trader's tips as you would for the double tops, or double bottoms, respectively. As shown in Figure 2.10, in a triple bottom formation, the bottoms have about the same amplitude. A parallel line (the neckline) is drawn against the line connecting the three bottoms (B, D, and F). As a support line, the neckline is broken at point A. It turns into a strong resistance for price levels C and E but eventually fails at point G. The resistance line turns into a strong support line, which holds the market backlash at point H. The price objective is at level I, which is the average length of the triple bottom formation, as measured from point D.

Market examples of triple tops and triple bottoms are presented in Figures 2.11 and 2.12.

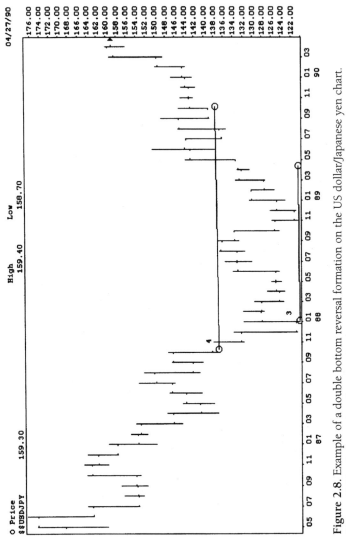

Figure 2.8. Example of a double bottom reversal formation on the US dollar/Japanese yen chart. (*Source:* Bridge Information Systems, Inc.)

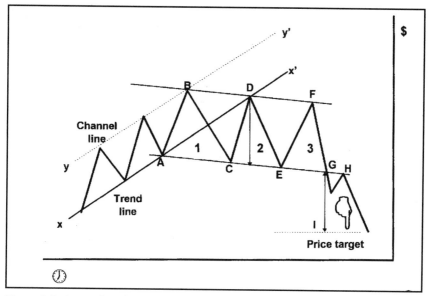

Figure 2.9. A typical triple top formation.

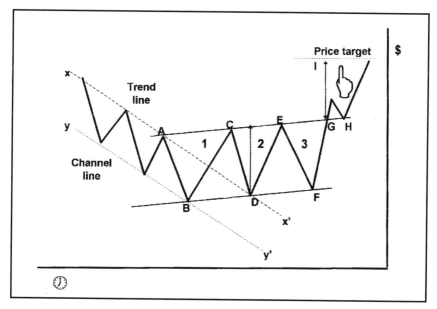

Figure 2.10. A typical triple bottom formation.

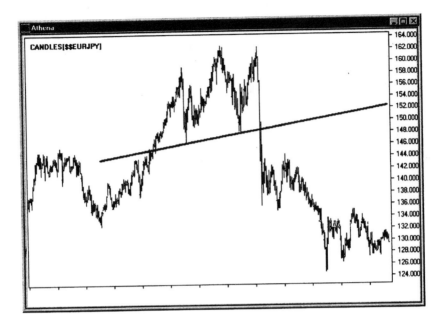

Figure 2.11. Example of a triple top reversal formation on the euro/Japanese yen chart. (*Source:* Bridge Information Systems, Inc.)

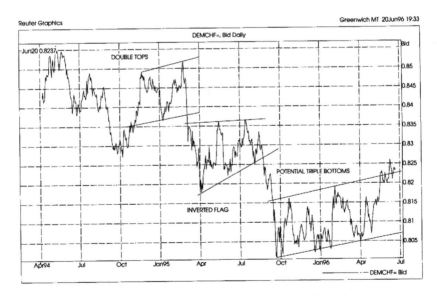

Figure 2.12. A potential triple bottom formation in the Swiss franc futures. (*Source:* Reuters.)

V-Formation (Spike)

The *V-formations,* or *spikes,* are an unusual type of major trend reversal formation. In all the other patterns discussed so far in this chapter, the market attempted to return the currencies to the trend after they broke out of it. Spike reversals lack this consolidation segment following the breakout. They simply switch from one trend to the opposite one without warning. V-formations are naturally characterized by heavy volume at and soon after the sudden switch because the outstanding positions opened during the original trend must be liquidated and/or reversed.

Figure 2.13 shows you the structure of the V-bottom formation. The bar at the bottom of the chart, which marks the low of the down trend, can be isolated from the range of either the previous or the following bar. (This possible development is discussed in detail in Chapter 4.) Both trends are sharp, denoting the certainty of the market in the direction of the currency. In the original down trend, the psychology of the market consists of positive thoughts, possibly backed by weak fundamentals for the specific currency.

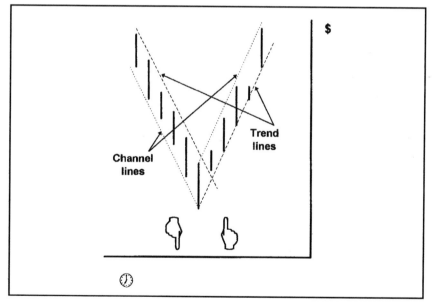

Figure 2.13. A typical V-bottom formation.

In the second trend, it is more difficult to tell how the market feels. Undoubtedly, a major event triggered an unusual shift in some traders' behavior. At the same time, other players seem to be uncertain about the significance of the reversal. Some may become convinced later and cover their short positions at a loss, while others may remain unconvinced but are forced to cut their losses anyway. Altogether, these changes in the market psychology generate a continuous currency buying process that extends both the duration and significance.

Figure 2.14 illustrates a V-bottom reversal formation on the US dollar/Japanese yen daily chart.

Figures 2.15 and 2.16 describe and provide an example of the V-top reversal formation.

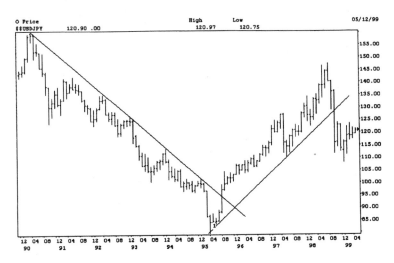

Figure 2.14. Example of a V-bottom reversal formation on the US dollar/Japanese yen chart. (*Source:* Bridge Information Systems, Inc.)

Rounded Top and Bottom

Rounded tops and bottoms (see Figures 2.17 and 2.18), also known as *saucers,* are infrequent trend reversal chart patterns. They reflect a gradual change in the market's direction and its indecision at the end of a trend. Will the original trend continue or has it bottomed out/topped out? As the bulls and bears tug in two directions, trading is slow. Knowing when the formation is indeed completed is impossible, and you have no trigger for the rebound. Such peculiarities make these trend reversal formations unpopular among traders. Like any other consolidation pattern, the longer the rounded top or bottom takes to complete itself, the higher the likelihood of a sharp price move in the new direction.

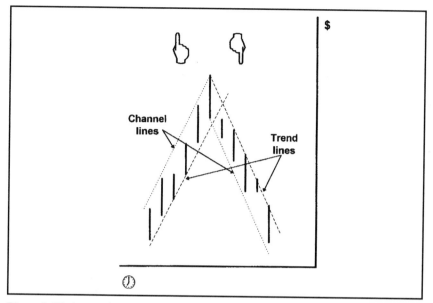

Figure 2.15. A typical V-top reversal formation.

Figure 2.16. A typical V-top reversal formation in the US dollar/Japanese yen weekly chart. (*Source:* CQG. ©Copyright CQG INC.)

Figure 2.17 shows you the structure of the rounded bottom pattern, and Figure 2.19 depicts the rounded top formation. Figures 2.18 and 2.20 show you market examples of these reversal formations.

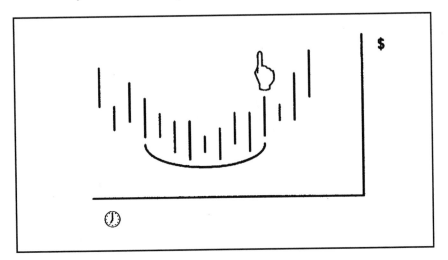

Figure 2.17. A typical rounded bottom (saucer) formation.

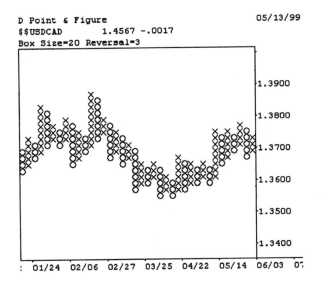

Figure 2.18. Example of a rounded bottom reversal pattern on the US dollar/Canadian dollar point and figure chart. (*Source:* Bridge Information Systems, Inc.)

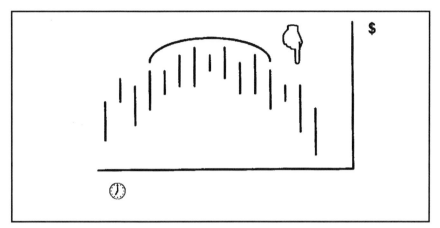

Figure 2.19. A typical rounded top formation.

Diamond Formation

The *diamond* formation is a minor reversal pattern that tends to occur at the top of the trend. The price activity can be outlined by a shape resembling a diamond. See Figure 2.21. The combination of divergent and convergent support and resistance lines is closely mimicked by the increase and decrease in trading volume. Upon the breakout, volume picks up substantially. The price target is the height of the diamond, measured from the breakout point.

This reversal formation is not popular because it has a high rate of false breakouts. It is, however, mercifully rare in the currency markets.

An example of a diamond reversal formation in the sterling/US dollar market is provided in Figure 2.22.

Rectangle Reversal

The *bearish rectangle* reversal formation shows a common feature of market psychology at the end of an up trend: consolidation before reversal. The consolidation area is encompassed by a rectangle; hence its name. See Figure 2-23. The bearish breakout from a rectangle yields a price objective equal in size to the height of the rectangle, and it is measured from the breakout point. The height of the rectangle in the figure is $A'B'$, and its price target is $B'B'$, measured from the trigger point B'.

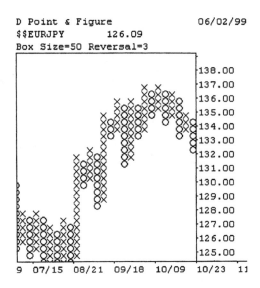

Figure 2.20. Example of a rounded top formation. (*Source:* Bridge Information Systems, Inc.)

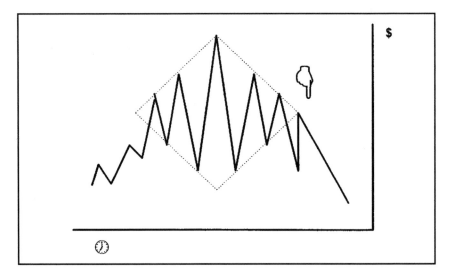

Figure 2.21. A typical diamond formation.

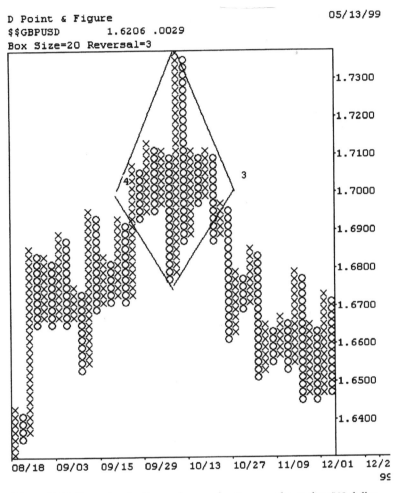

Figure 2.22. Example of a diamond reversal pattern on the sterling/US dollar point and figure chart. (*Source:* Bridge Information Systems, Inc.)

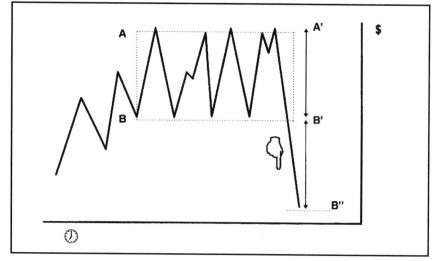

Figure 2.23. A typical bearish rectangle reversal formation.

The *bullish rectangle* reversal formation shows a consolidation at the end of a down trend. See Figure 2.24. The bullish breakout from a rectangle yields a price objective equal in size to the height of the rectangle, and it is measured from the breakout point. The height of this rectangle is $A'B'$, and its price target is $B'B''$, measured from the trigger point B'.

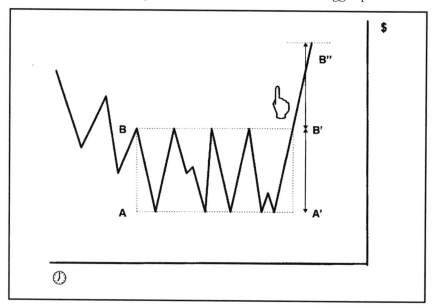

Figure 2.24. A typical bullish rectangle reversal formation.

Market examples of the bullish and bearish rectangles are illustrated in Figure 2.25 on the US dollar/Swiss franc chart.

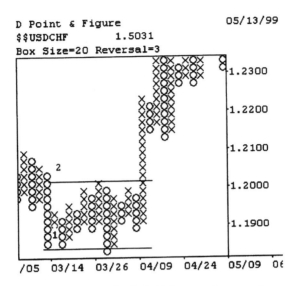

Figure 2.25. Example of a bullish rectangle reversal pattern on the US dollar/Swiss franc point and figure chart. (*Source:* Bridge Information Systems, Inc.)

CONTINUATION PATTERNS

In addition to the major trend reversal patterns, technical analysts also identify chart formations that reinforce trends, known as *continuation patterns*. These patterns reflect the "break" the market needs during a sharp trend. During these formations, some traders realize some or all of their profit, while others wonder whether the trend has already run its course. These activities, as a whole, generate consolidations that sometimes occur as neutral moves, but usually slope against the original trend. The consolidations are fairly short. The breakouts must naturally occur in the same direction as the original trend.

Price Structure

The most important continuation patterns are the:

1. Flag.
2. Pennant.
3. Triangle.
4. Wedge.
5. Continuation rectangle.

Flag Formation

The *flag* is a reliable trend continuation pattern that provides two vital signals: direction and price objective. This formation consists of a brief consolidation period within a solid and steep upward or downward trend. The consolidation itself tends to be sloped in the opposite direction to the slope of the original trend, or it can be simply flat. The consolidation is bordered by a support line and a resistance line, which are either parallel to each other (as in a parallelogram) or very gently converging, making the pattern look like a flag. The original trend is the *flagpole*. Once the currency resumes the trend by breaking out of the consolidation, the price objective is the total length of the flagpole, measured from the breakout price level.

Bullish Flag

If the original trend is up, the formation is called a *bullish flag*. In Figure 3.1, the original trend is sharply up. The flagpole is measured between the price levels *A* and *B*. The consolidation period occurs between the support line *DE* and the resistance line *BC*. After the market penetrates the resistance line at point *C,* the trend resumes its rally, with the price objective *F,* measured from *C*. The price target *CF* is equal to the flagpole's entire length *AB*.

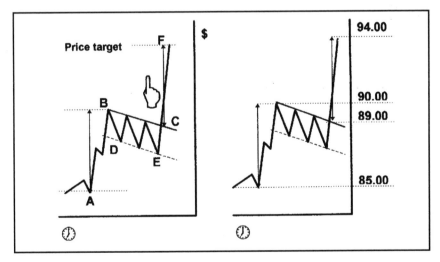

Figure 3.1. A typical bullish flag.

In the numerical example, the height of the flagpole is measured as the difference between 90.00 and 85.00, or 500 pips. Once the resistance line is broken at 89.00, the price target is 94.00, or 500 pips from breakout point at 89.00.

Figure 3.2 gives you a market example of a bullish flag in the Australian dollar/US dollar chart.

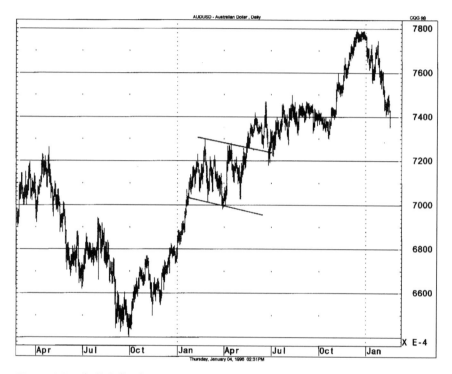

Figure 3.2. A bullish flag formation in Australian dollar/US dollar. (*Source:* CQG. ©Copyright CQG INC.)

Bearish Flag

Figure 3.3 displays the price development of a *bearish flag* formation. In this case the original trend is bearish. The flagpole is measured between the *A* and *B* price levels. The consolidation period occurs between the resistance line *DE* and the support line *BC*. When the market penetrates the support line at point *C*, the trend resumes its fall, with the price objective *F*, measured from breakout point *C*. The price target *CF* has about the same amplitude as the flagpole's length.

In the numerical example, the height of the flagpole is measured as the difference between 1.5000 and 1.4500, or 500 pips. Once the support line is broken at 1.4600, the price target is 1.4100, or 500 pips below 1.4600.

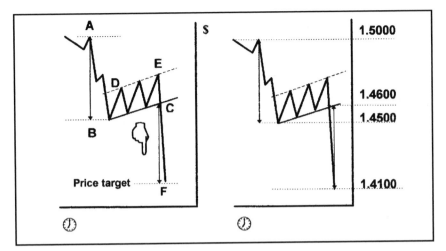

Figure 3.3. A typical bearish flag formation.

Figure 3.4 shows you the bearish or inverted flag pattern in the US dollar/Japanese yen chart.

As you can see from Figures 3.1 through 3.4, flags are reliable formations. They tend to develop over periods that vary from days to weeks. However, due to false breakouts, the support and resistance lines delimiting the flag may have to be adjusted several times.

Pennant Formation

Pennants are closely related to flags, and the same principles apply. Their sole difference from flags is that the consolidation area more closely resembles a pennant, since the support and resistance lines converge. In terms of market psychology, the triangular pennant reflects not only consolidation, but also a lack of liquidity. The currency price fluctuations decline rapidly as the market looks for a new direction. Traders must be careful when determining the direction and the validity of the breakout in order to avoid costly mistakes.

Bullish Pennant

If the original trend is bullish, then the chart pattern is a *bullish pennant*. In Figure 3.5, the pennant pole is *AB*. The pennant-shaped consolidation is framed by points *C, B,* and *D*. When the market breaks through the resistance line *BC* on significant volume, the price objective becomes *E*.

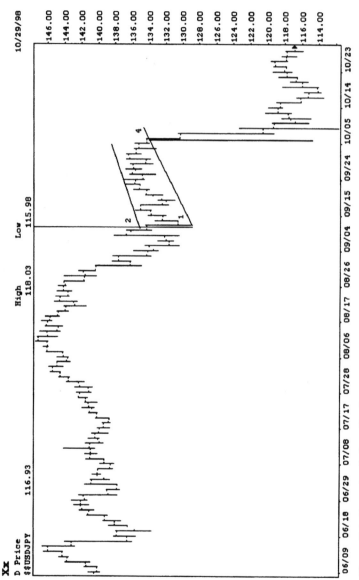

Figure 3.4. Example of a bear flag formation on the US dollar/Japanese yen chart. (*Source:* Bridge Information Systems, Inc.)

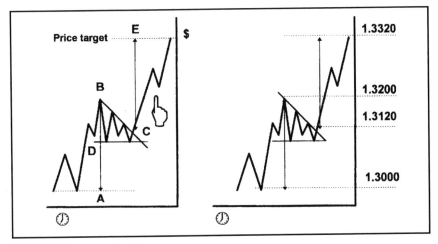

Figure 3.5. A typical bullish pennant.

The amplitude of the target price is *C* to *E,* and it is equal to the pennant pole *A* to *B.* The price target measurement starts from the breakout point.

In the numerical example, the height of the pennant pole is measured as the difference between 1.3200 and 1.3000, or 200 pips. Once the resistance line is broken at 1.3120 on high volume, the price target is 1.3320, or 200 pips from 1.3120.

Figure 3.6 illustrates a bullish pennant in the currency market.

Bearish Pennant

If the original trend is going down, the formation is a *bearish,* or *inverted, pennant.* In Figure 3.7, the pennant pole is *AB.* The pennant-shaped consolidation is framed by *C, B,* and *D.* When the market breaks through the support line *BC,* the objective price is *E.* The amplitude of the target price is *CE,* and it is equal to the pennant pole *AB.* The price target measurement starts from the breakout point.

In the numerical example, the height of the pennant pole is measured as the difference between 1.7400 and 1.7200, or 200 pips. Once the support line is broken on heavy volume at 1.7270, the price target becomes 1.7070, or 200 pips from 1.7270.

As you can see, pennants are as reliable as flags. They may develop in different time frames, from days to weeks. Again, however, the support and resistance lines may have to be adjusted to avoid the danger of false breakouts. Figure 3.8 displays a market example of a bearish pennant in the Australian dollar/US dollar market.

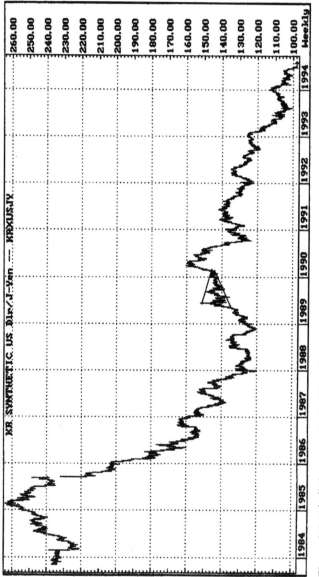

Figure 3.6. A bullish pennant in the US dollar/Japanese yen. (*Source:* Bridge Information Systems, Inc.)

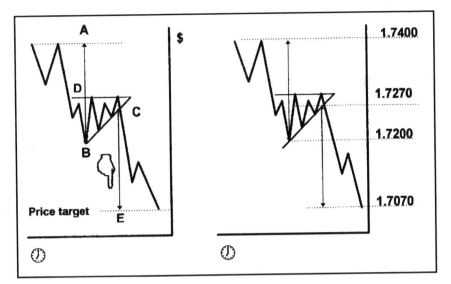

Figure 3.7. A typical bearish pennant.

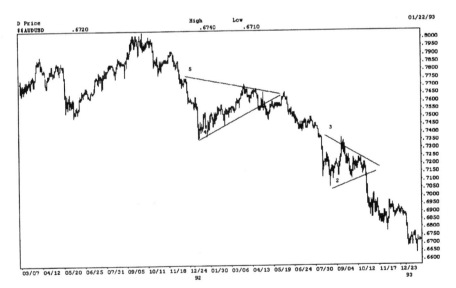

Figure 3.8. Examples of bearish pennant formations on the Australian dollar/US dollar chart. (*Source:* Bridge Information Systems, Inc.)

Triangle Formation

Triangles can be visualized as pennants with no poles. This means that the price objectives are smaller than those of pennants. It also means that, since it takes fewer prices to form this pattern, it occurs more often. There are four types of triangles: symmetrical, ascending, descending, and expanding (broadening).

Symmetrical Triangle

A *symmetrical triangle* is delimited by symmetrically converging support and resistance lines, defined by at least four significant points. In Figure 3.9, the support line is defined by points *B* and *C*, the resistance line by points *A* and *C*. The price target of a triangle can be measured in two ways: (1) as an extrapolation of the width of the base *AB* of the triangle from the breakout point *D*, or (2) as the intersection of the price line with the line *AE*, which is parallel to the support line *BC*. These two methods may lead to different price objectives.

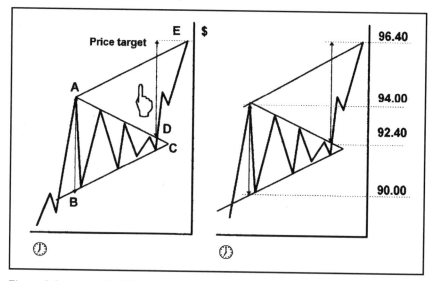

Figure 3.9. A typical bullish symmetrical triangle.

A breakout may occur on either side.

In a *bullish symmetrical triangle*, the breakout must occur in the same direction as the previous bullish trend to qualify the formation as a continuation pattern. Trading volume visibly decreases toward the tip of the triangle, suggesting either ambivalence in the market or a temporary balance between supply and demand. The breakout must be accompanied by a rise in volume.

In the numerical example, the size of the bullish triangle base is measured as the difference between 94.00 and 90.00, or 400 pips. The price objective is 400 pips above 92.40 at 96.40.

Figure 3.10 shows a market example of a bullish symmetrical triangle in the IMM Japanese yen futures.

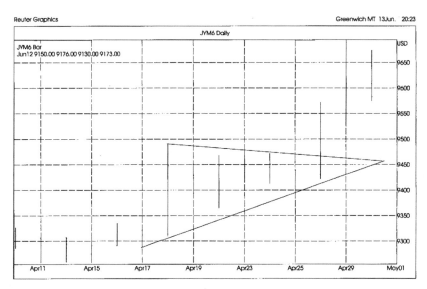

Figure 3.10. A typical bullish symmetrical triangle in the IMM Japanese yen futures. (*Source:* Reuters.)

Figure 3.11 presents a typical *bearish symmetrical triangle*. The consolidation area is enclosed by *ABC*. During a bearish trend, the market sells aggressively through the support *AB* and breaks out downward at point *D*. The size of the triangle is *CA* and the price objective is *DE*.

In the numerical example, the height of the triangle is measured as the difference between 94.00 and 90.00 (400 pips). The price objective is also 400 pips down from 91.50: 87.50.

A currency market example of a bearish symmetrical triangle is presented in Figure 3.12.

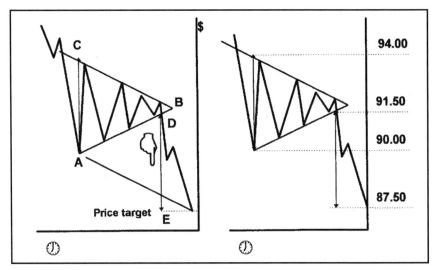

Figure 3.11. A typical bearish symmetrical triangle.

Ascending Triangle

The *ascending triangle* is a continuation formation that is delimited by a flat resistance line and an upward sloping support line. Ascending triangles can be bullish and bearish.

In Figure 3.13, the *bullish ascending triangle* follows an up trend. The formation suggests that the demand remains stronger than the supply, despite the consolidation. The pattern is confirmed when the breakout occurs on the up side on good volume. The price target is equal to the width of the base of the triangle as measured from the breakout point. Trading volume decreases steadily toward the tip of the triangle, but increases rapidly on the breakout.

In the figure, the resistance line *AB* is flat. The converging support line *CB* slops upward. The price objective *D* is the width of the base of the triangle *AC,* measured above the resistance line from the breakout point.

In the numerical example, the height of the ascending triangle is measured as the difference between 105.00 and 103.00. The price objective is 107.00, or 200 above 105.00.

Figure 3.14 presents a market example of a bullish ascending triangle in the US dollar/Japanese yen chart.

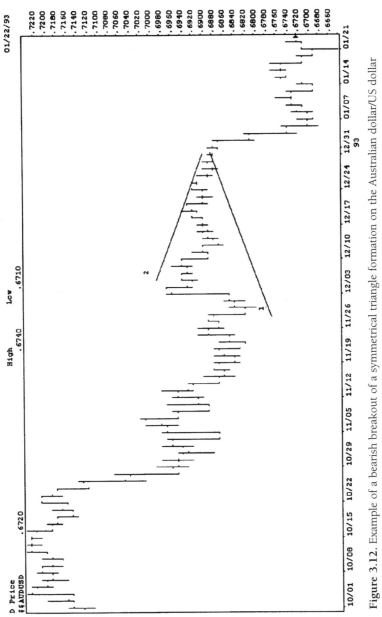

Figure 3.12. Example of a bearish breakout of a symmetrical triangle formation on the Australian dollar/US dollar chart. (*Source:* Bridge Information Systems, Inc.)

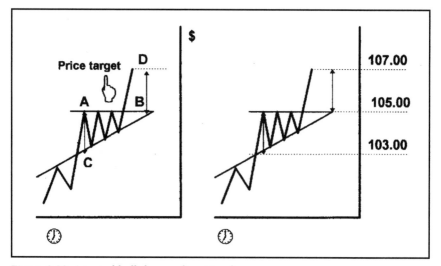

Figure 3.13. A typical bullish ascending triangle.

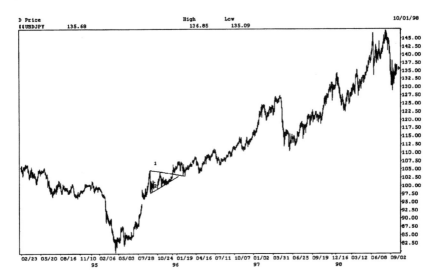

Figure 3.14. Example of a bullish breakout of an ascending triangle formation on the US dollar/Japanese yen chart. (*Source:* Bridge Information Systems, Inc.)

As illustrated in Figure 3.15, the *bearish ascending triangle* continues a down trend. The formation suggests that the demand remains weaker than the supply, despite the consolidation. The bearish ascending triangle is confirmed when the breakout occurs on the down side on good vol-

ume. The price target equals the width of the base of the triangle as measured from the breakout point. Trading volume decreases steadily toward
the tip of the triangle, but increases rapidly on the breakout.

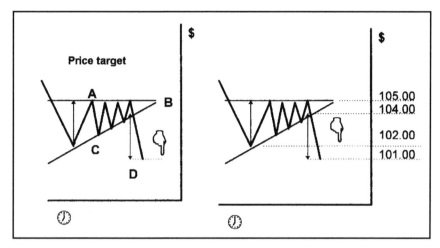

Figure 3.15. A typical bearish ascending triangle.

In the figure, the flat resistance line is defined by points *A* and *B*.
The converging support line *CB* slopes upward. The price objective is the
width of the base of the triangle *AC*, measured below the support line
from the breakout point.

In the numerical example, the height of the ascending triangle is
measured as the difference between 105.00 and 102.00. The price objective is 101.00, or 300 below 104.00.

Descending Triangle

The *descending triangle,* the mirror image of the ascending triangle, is
characterized by a flat support line and a downward sloping resistance
line. This continuation pattern may be bullish or bearish.

The *bullish descending triangle* occurs in an up trend, and it consists
of a flat lower support line and a downward sloping resistance line (see
Figure 3.16). The pattern suggests that the demand is larger than the supply, but the currency is expected to breakout on the upside. Despite the
buying interest in the market, the resilience of the resistance line negatively affects volume. However, the upward breakout must occur on sig-

nificant volume to validate the formation. Like the other triangles, the descending pattern provides a price objective that is calculated by measuring the height of the triangle base and then extrapolating it from the breakout point.

In Figure 3.16, the horizontal support line is defined by points *A* and *C*. The converging resistance line *BC* slopes downward. The price objective *CD* equals the height of the triangle *AB*, measured from the breakout point through the resistance line.

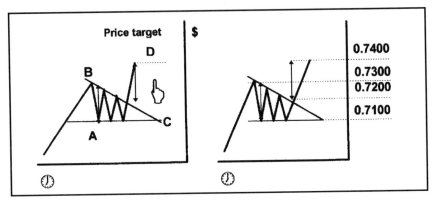

Figure 3.16. A typical bullish descending triangle.

In the numerical example, the height of the descending triangle measures 200 pips, the difference between 0.7300 and 0.7100. The price objective is 0.7400, or 200 pips above 0.7200.

A *bearish descending triangle* continues a down trend, and it consists of a flat lower support line and a downward sloping resistance line (see Figure 3.17). This pattern suggests that the supply is larger than the demand, with the currency expected to breakout on the downside. Despite the selling interest in the market, the resilience of the support level negatively affects volume, but the downward breakout must be accompanied by heavy volume to validate the formation. The descending triangle's price objective is calculated by measuring the height of the triangle base and then extending it from the breakout point.

In Figure 3.17, the horizontal support line is defined by points *A* and *C*. The converging resistance line *BC* slopes downward. The price objective *CD* equals the height of the triangle *AB*, measured from the breakout point through the support line.

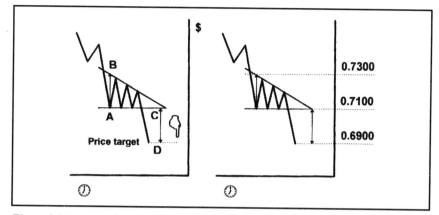

Figure 3.17. A typical bearish descending triangle.

In the numerical example, the height of the descending triangle measures 200 pips, the difference between 0.7300 and 0.7100. The price objective is 0.6900, the 200-pip difference between 0.7100 and 0.6900.

Figure 3.18 displays a market example of a bearish descending triangle in the US dollar/Canadian dollar.

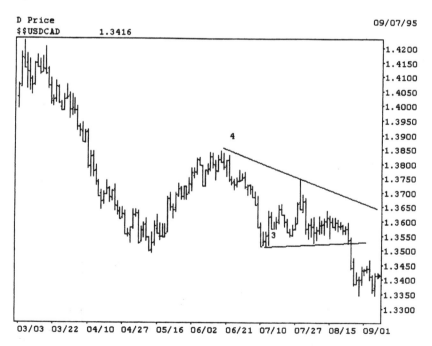

Figure 3.18. A descending triangle in the US dollar/Canadian dollar. (*Source:* Bridge Information Systems, Inc.)

Expanding, or Broadening, Triangle

The *expanding, or broadening, triangle* is a fairly unusual pattern. It is a horizontal mirror image of a symmetrical triangle, but the tip of the triangle, not its base, is next to the original trend. Volume follows the horizontal mirror image switch and increases steadily as the chart formation develops. Expanding triangles may be bullish or bearish.

In Figure 3.19, the *bullish expanding triangle* consists of the divergent support line *AB* and the resistance line *AC*. The price objective *CD* is calculated as the height *BC* of the triangle, measured from the breakout point *C*.

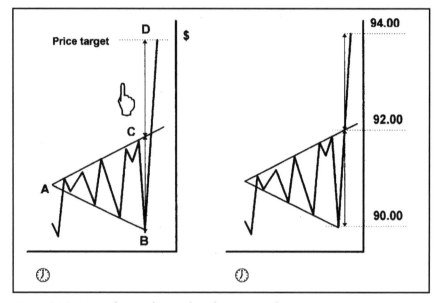

Figure 3.19. A typical expanding, or broadening, triangle.

In the numerical example, the height of the triangle is 200 pips (92.00 – 90.00), and the price objective is 94.00, the 200-pip difference between 94.00 and 92.00, measured from 92.00.

Figure 3.20 gives you a market example of an expanding, or broadening, triangle in the Australian dollar/US dollar chart.

A *bearish expanding triangle* continues a down trend. In Figure 3.21, the pattern consists of the divergent support line *AC* and a resistance line *AB*. The price objective *CD* is calculated as the height *BC* of the triangle, measured from the breakout point *C*.

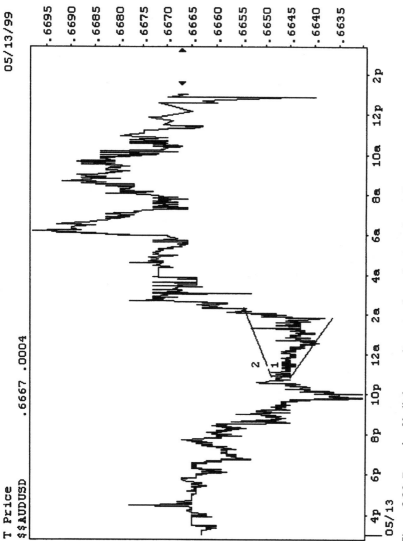

Figure 3.20. Example of bullish expanding triangle on the Australian dollar/US dollar chart.
(*Source:* Bridge Information Systems, Inc.)

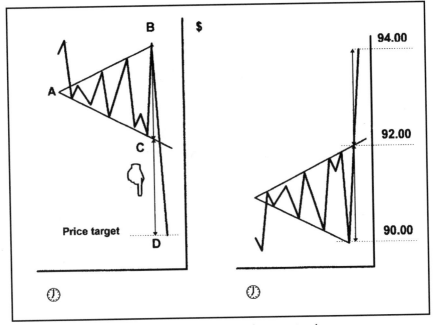

Figure 3.21. A typical bearish expanding, or broadening, triangle.

In the numerical example, the height of the triangle is 200 pips (92.00 – 90.00) in USD/JPY and the price objective is 94.00, as the 200-pip difference between 94.00 and 92.00, measured from 92.00.

Figure 3.22 gives you a market example of a bearish expanding, or broadening, triangle in the US dollar/Canadian dollar chart.

Wedge Formation

The *wedge formation* is a close relative of the triangle formation. In Figure 3.23, the wedge slopes markedly against its original trend, and the break-out occurs in the direction opposite to its slope. The signal from the wedge formation is only for direction. There are no reliable price objectives, other than the market's attempt to return to where the wedge began. Depending on the trend direction, there are two types of wedges: falling and rising. A falling wedge occurs in an up trend, and a rising wedge occurs in a down trend. In either case, volume must be significant on breakouts.

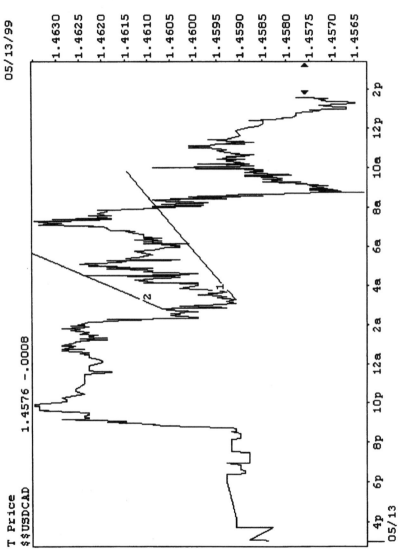

Figure 3.22. Example of bearish expanding triangle on the US dollar/Canadian dollar chart. (*Source:* Bridge Information Systems, Inc.)

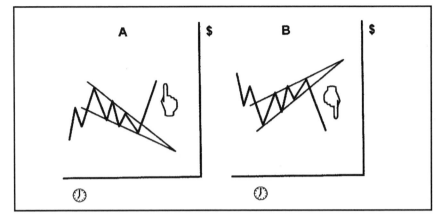

Figure 3.23. A typical rising wedge (A) and a typical falling wedge (B).

Figures 3.24 and 3.25 show you practical examples of wedges in the currency market.

Continuation Rectangle Formation

Also known as a *trading range* (or *congestion*), the *rectangle formation* reflects a consolidation period similar to a flag without a pole. Upon breakout, the original trend is likely to continue. Failure to do so changes the formation from a continuation to a reversal pattern. This pattern is easy to spot, since it can be considered a minor sideways trend. If it occurs within an up trend and the breakout occurs on the up side, it is called a *bullish rectangle.* The price objective is the height of the rectangle measured from the breakout point through the resistance line. Volume should give further indication at, or close to, the points tangential to the support and resistance levels. As usual, heavy volume predicts, or substantially increases, the chance of a breakout.

In the bullish rectangle in Figure 3.26, you can see the currency moving between well-defined, flat support and resistance levels. A valid breakout may occur on either side of this consolidation period. The price target *CE* is equal to the height *AC* of the rectangle, measured from the breakout point through the resistance line *CD*.

In the numerical example, the height of the rectangle equals 200 pips (1.4800 – 1.4600), and the price objective is 1.5000, the 200-pip difference between 1.5000 and 1.4800, measured from the resistance line at 1.4800.

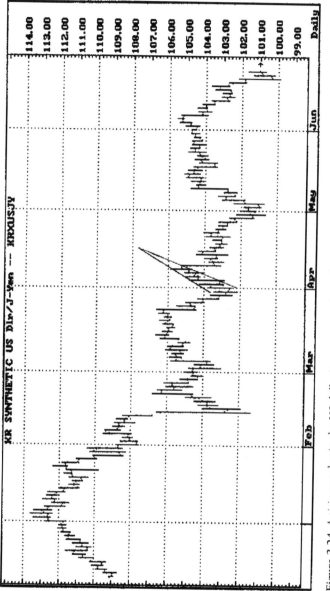

Figure 3.24. A rising wedge in the US dollar/Japanese yen chart. (*Source:* Bridge Information Systems, Inc.)

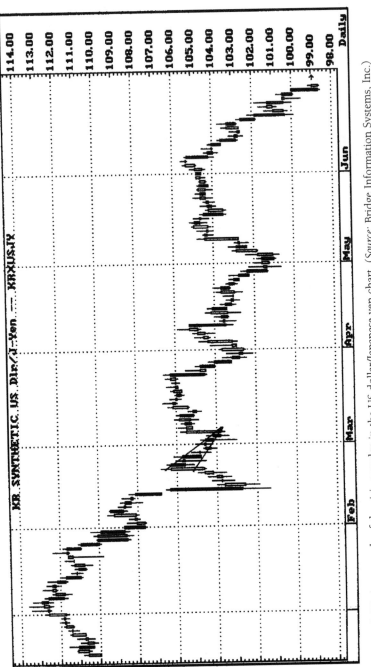

Figure 3.25. An example of the rising wedge in the US dollar/Japanese yen chart. (*Source:* Bridge Information Systems, Inc.)

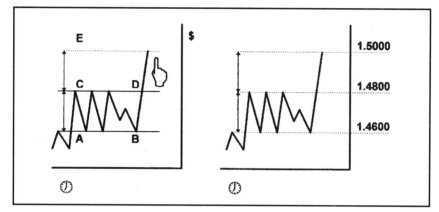

Figure 3.26. A typical bullish rectangle.

If the consolidation occurs within a down trend and the breakout occurs on the downside, the formation is called a *bearish rectangle*. Figure 3.27 shows you a currency consolidation between well-defined, flat support *CD* and resistance *AB* lines. A valid breakout may occur on either side of this consolidation period. The price objective *DE* is equal in size to the height *BD* of the rectangle, measured from the breakout point through the support line *CD*.

In the numerical example, the height of the bearish continuation rectangle is the 200-pip difference between 1.4800 and 1.4600. The price objective is 1.4400, the 200-pip difference between 1.4600 and 1.4400, measured from 1.4600.

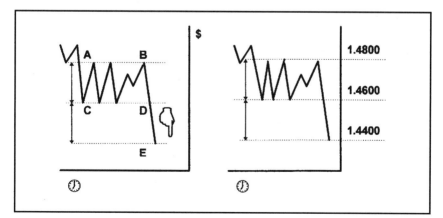

Figure 3.27. A typical bearish continuation rectangle.

PART 2

TYPES OF CHARTS

FORMATIONS UNIQUE TO BAR CHARTS FOR FUTURES

Unlike the cash market, currency futures do not trade actively on a 24-hour basis. This characteristic gives rise to specific trading signals that are profitable for both cash and futures traders. Since the bar chart is the chart of choice among currency futures traders, this chapter analyzes chart formations unique to bar charts on currency futures. These patterns are gaps, island reversals, and key reversals.

While line or point and figure charts are always continuous charts, bar charts (and candlestick charts, for that matter) may or may not be continuous. The bar chart is continuous for the currency cash market and may be discontinuous for the currency futures market. It is continuous for the cash market because every bar opens at the same price level at which the previous bar closed. For instance, if the daily (24 hours) bar chart for USD/JPY closes at 110.00 at 5:00 PM EDT today, then the bar for tomorrow must open at 110.00 at 5:01 PM EDT (see Figure 4.1).

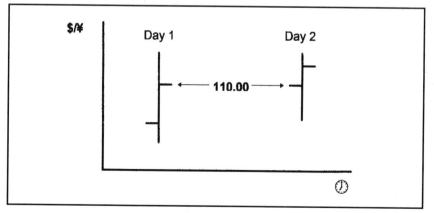

Figure 4.1. A bar chart structure in the USD/JPY spot market.

Figure 4.2 shows you a bar chart in the US dollar/Japanese yen cash market.

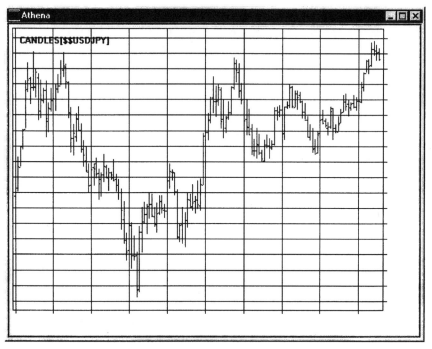

Figure 4.2. A bar chart of the US dollar/Japanese yen chart. (*Source:* Bridge Information Systems, Inc.)

Gaps

Exceptions may occur between a Friday's range and a Monday's opening price. If a significant event occurs over the weekend, when there is barely any currency trading, the market in New Zealand or Australia might open on a Monday outside Friday's range. An opening outside the previous day's or other period's range generates a *price gap*.

Price gaps, as plotted on bar charts, are very common in the currency futures market. Although currency futures may be traded around the clock, their markets are open for only about a third of the trading day. For instance, the largest currency futures market in the world, the Chicago IMM, is open for business 7:20 AM to 2:00 PM CDT. Since the cash market continues to trade around the clock, price gaps may occur between two days' price ranges in the futures market. In Figure 4.3 you can see two daily bars. The first bar has a high of 0.6580 and the second day, which opened higher than the first day's high, has a low of 0.6610. The price gap formed between the first day's high and the next day's low is 30 pips.

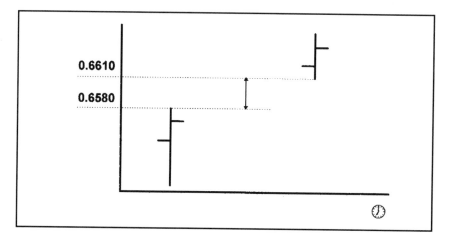

Figure 4.3. A price gap in the futures market.

Currency futures trading sessions are shorter than cash trading sessions and price gaps may occur on a daily basis, if the overnight activity exceeds the previous day's range. Although the following patterns are applicable to both bar and candlestick charts, we will analyze each type of chart separately.

A market example of a bar chart in the IMM British pound futures market is presented in Figure 4.4.

There are four types of gaps: *common, breakaway, runaway,* and *exhaustion.*

Common Gaps

Common gaps have the least technical significance of all the types of gaps. They do not indicate a trend start, continuation, reversal, or even a general direction of the currency other than in the very short term. Figure 4.5 shows examples of common gaps in the Japanese yen futures. Gaps *B* and *C* are obvious. These price gaps were closed later. Price gap *A* is less obvious because it was closed on the same day.

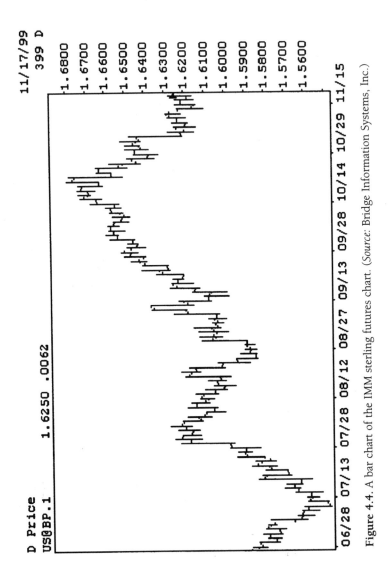

Figure 4.4. A bar chart of the IMM sterling futures chart. (*Source:* Bridge Information Systems, Inc.)

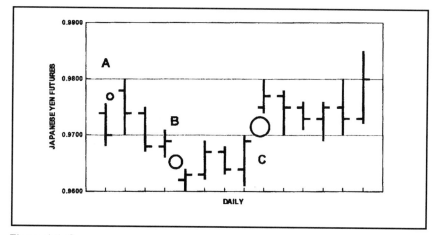

Figure 4.5. Common gaps.

Common gaps tend to occur in relatively quiet periods or in illiquid markets. When price gaps occur in illiquid markets, such as distant currency futures expiration dates, they must be completely ignored. The entries for distant expiration dates in currency futures are made only on a closing basis, and they do not reflect any trading activity. Never trade in an illiquid market because getting out of it is very difficult and expensive. When gaps occur within regular trading ranges, the word on the street has been that, *"Gaps must be filled."*

Common gaps are short term. When currency futures open higher than yesterday's high, they are quickly sold, targeting the level of the previous day's high. Should the selloff fail, the market turns on a dime, reversing the short-term position. For instance, if the previous day's range is 0.7030–0.7100 and today's opening price is 0.7130, traders tend to sell the futures contracts from 0.7130 to 0.7100 to fill the price gap. If they cannot reach 0.7100, then the market quickly turns to buying the futures contracts. The opposite is true if the market opens lower than yesterday's low, as traders buy the futures contracts to fill the price gap between the current lower opening price and yesterday's low. For instance, if the previous day's range is 0.7030–0.7100 and today's opening price is 0.7000, traders tend to buy the futures contracts from 0.7000 to 0.7030 to fill the price gap. If they cannot reach 0.7030, then the market quickly turns to selling the futures contracts. Although professional traders agree that the common gap lacks technical significance, it is not always clear whether they are dealing with a common gap or another type of gap. Until more information becomes available, the "gap filling" continues.

Figure 4.6 illustrates common gaps in the Australian dollar IMM futures market.

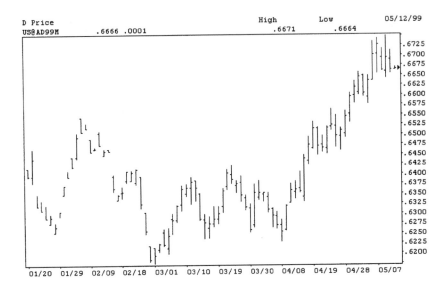

Figure 4.6. Examples of common gaps on the Australian dollar IMM futures. (*Source:* Bridge Information Systems, Inc.)

Breakaway Gaps

Breakaway gaps occur at the beginning of a new trend, usually at the end of long consolidation periods. They may also appear after the completion of some chart formations that tend to act as short-term consolidations. Breakaway gaps signify a brisk change in trading sentiment, and they occur on increasingly heavy trading. Traders are understandably frustrated by consolidations, which are rarely profitable. Therefore, a breakout from the slow lane is embraced with optimism by the profit-hungry traders. The price takes a secondary place to participation. As always, naysayers follow the initial breakout. Sooner rather than later, the pessimists have no choice but to join the new move, thus creating more volume.

Breakaway gaps are not likely to be filled during the breakout and for the duration of the subsequent move. In time, they may be filled during a new move on the opposite side.

In Figure 4.7, the currency futures trades sideways in a 100-pip range between 0.6550 and 0.6690 for a period of time. A price gap between 0.6690 and 0.6730 signals the breakaway from the range. The market is likely to aggressively buy currency futures.

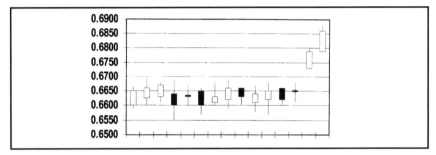

Figure 4.7. A typical breakaway gap.

Figure 4.8 illustrates a breakaway gap in the Canadian dollar IMM futures market.

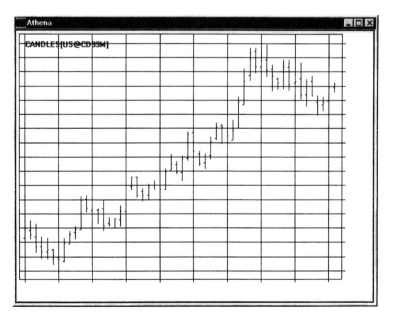

Figure 4.8. Example of a breakaway gap on Canadian dollar IMM futures. (*Source:* Bridge Information Systems, Inc.)

Trading Signals for Breakaway Gaps

1. A breakaway gap provides the price direction.
2. There is no price objective.
3. Increasing demand for a currency ensures a solid move on good volume in the foreseeable future.

Runaway, or Measurement, Gaps

From a technical point of view, *runaway,* or *measurement, gaps* are special gaps that occur within solid trends. They are known as measurement gaps because they tend to occur about midway through the life of a trend. Thus, if you measure the total range of the previous trend and extrapolate it from the measurement gap, you can identify the end of the trend and your price objective. Since the velocity of the move should be similar on both sides of the gap, you also have a time frame for the duration of the trend.

In Figure 4.9, you can identify an up trend measuring 250 pips, between 0.6590 and 0.6840, and a 40-pip runaway gap, which occurs at the top of an up trend between 0.6590 and 0.6880. In this case, measure the previous 200-pip range from the top of the gap (0.6880) and calculate the price objective as 0.7130.

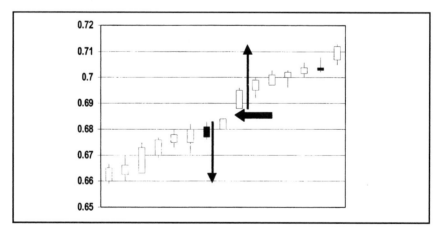

Figure 4.9. A typical measurement (runaway) gap.

Figure 4.10 shows a runaway gap in the euro IMM futures.

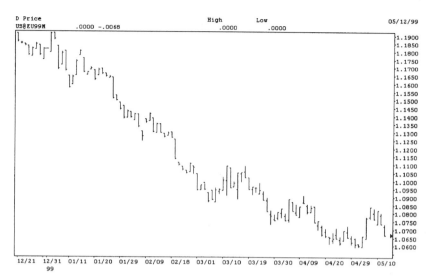

Figure 4.10. Example of a potential measurement gap on the euro IMM futures. (*Source:* Bridge Information Systems, Inc.)

Trading Signals for Runaway Gaps

1. The runaway, or measurement, gap provides the direction of the market. As a continuation pattern, this type of gap confirms the health and the velocity of the trend.

2. Volume is good because traders like trends, and confirmed trends attract more optimism and capital.

3. This is the only type of gap that also provides a price objective and a time frame. These characteristics are also useful for developing hedging strategies.

Exhaustion Gaps

Exhaustion gaps may occur at the top or bottom of a V-formation, when trends change direction in an uncharacteristically quick manner. There is no consolidation next to the broken trend line: The trend reversal is very sharp.

In the left half of Figure 4.11, British pound futures rally steadily from 1.6900, reaching a high of 1.7350. The following day the market opens higher with a total price gap of 50 pips between 1.7350 and 1.7400. The market is unable to close the gap. A gap like this, occurring through a bullish move, looks a lot like a measurement gap. So traders buy the currency and stay long overnight on that assumption. The following day the market opens below the previous low, generating a second gap. Confused traders try to close the gap by buying even more currency in a falling market. As it becomes clear that the gap cannot be filled, it also becomes clear that they are dealing with an exhaustion gap. Long currency futures are closed and reversed; volume increases considerably and stays heavy in the medium term. A new trend is generated.

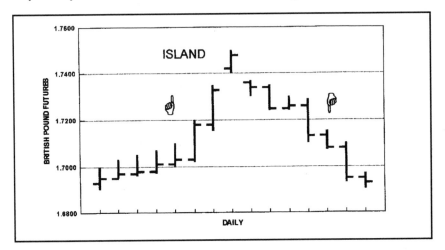

Figure 4.11. The price structure of an exhaustion gap.

If the second gap is filled or does not even occur, the trading signal remains the same. Traders do not have to get caught badly in this exhaustion gap. A sudden trend reversal is unlikely to occur in an information void. Some sort of identifiable event triggers the move—maybe a government fall or a massive and well-timed central bank intervention. Therefore, traders should at least be warned.

Although V-formations are not specific to currency futures bar charts, exhaustion gaps are. The presence of this type of gap greatly improves traders' confidence in the next price move.

Figure 4.12 shows a market example of an exhaustion gap in the Canadian dollar IMM futures.

Trading Signals for Exhaustion Gaps

1. The exhaustion gap provides direction of the market.

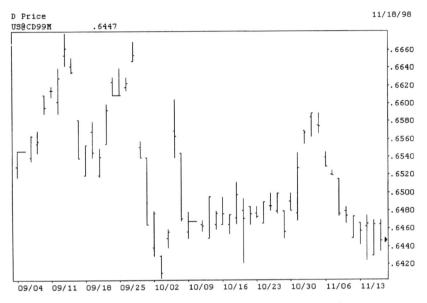

Figure 4.12. Examples of exhaustion gaps on Canadian dollar IMM futures. (*Source:* Bridge Information Systems, Inc.)

2. Volume is good. Traders like trends and a new trend attracts activity in the medium term.
3. There is no price objective.

Island Reversal

Island reversals are a direct result of exhaustion gaps. These bars occur at the tips of V-formations, where they are separated from the before and after ranges as isolated ranges, just like islands. When they appear in connection with the exhaustion gaps, they signal a very strong reversal. Island reversals may last for one or more days (see Figure 4.11).

Key Reversal Day

In a bullish market, a *reversal day* consists of a trading range that reaches a new high but closes lower than the previous day's high. Figure 4.13 shows an example of a vally to 2.0950. The last bar on the chart makes a new high at 2.1000 but manages to close at 2.0925, within the previous day's range of 2.0900 to 2.0950. The pattern should reflect up side exhaustion, but this reversal signal has a high rate of failure.

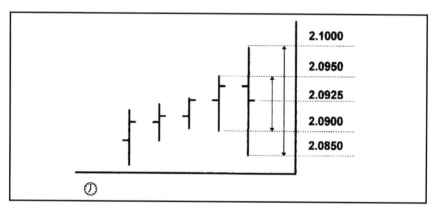

Figure 4.13. The price structure of reversal day.

A *key reversal day* occurs on the bar chart when the daily price range of the reversal day fully engulfs the previous day's range, eventually closing outside the previous day's range (see Figure 4.14). The reversal signal is more powerful than that of a reversal day.

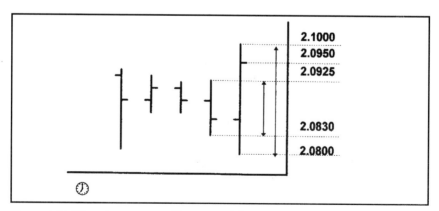

Figure 4.14. The price structure of key reversal day.

Key reversals may be slow to trigger a steep price action. Markets are very volatile on heavy volume on the key reversal day itself. Unless there is a strong trigger, traders need time to digest the activity of that day. Consolidation on the following day or for several days is common. Reversals and key reversals are not unique to currency futures bar charts. As long as no price gap is involved, these patterns also occur in the cash markets.

A key reversal day example is illustrated in Figure 4.15.

Figure 4.15. Example of a key reversal on the US dollar/Japanese yen chart. (*Source:* Bridge Information Systems, Inc.)

The Four-Week Rule

Richard Donchian devised a basic, yet powerful tool known as the *four-week rule*. This rule suggests that you should go long when the currency exceeds the highs of the previous four trading weeks. Conversely, you should sell when the currency falls below the lowest price of the previous four business weeks.

Upthrust and Spring

An *upthrust* is a bull trap that leads to trend reversal. In this pattern, the currency exceeds a previous high during the day and closes at a new high. There is a limited following, even when the market does not reverse immediately. The upthrust is confirmed when the market closes below the pivot point and toward the bottom of the daily range. Figure 4.16 illustrates a basic upthrust.

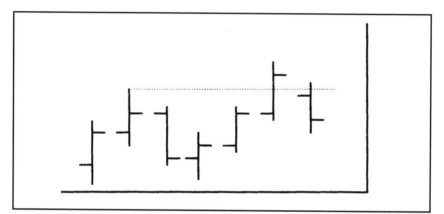

Figure 4.16. A basic upthrust.

A *spring* is a bear trap. See Figure 4.17. In this case, the currency must fall and close below a significant previous low. Again, the following is limited for one or more days. The market reverses and closes above the original low price.

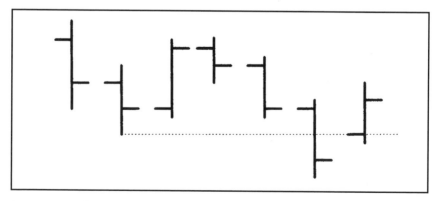

Figure 4.17. A basic spring reversal pattern.

POINT AND FIGURE CHARTING

Line and bar charts require two factors for plotting: prices and time periods. The point and figure chart is different from the previous three charts in that it needs only prices for analysis, not time periods. A point and figure chart also looks different from other charts and purposefully disregards some price activities. These idiosyncrasies provide easier-to-spot trading signals, heighten confidence, and increase profitability.

Price Structure

Toward the end of the nineteenth century, a new type of charting was developed: *the book method*. Under this system, if Charles Dow were transported forward in time about 100 years, he would enter the commodity prices in their original numerical format. For example, the prices in the USD/JPY would be recorded as shown in Figure 5.1. Entering, or figuring, the prices manually as in Figure 5.1A was obviously not popular. Even in the "good old days," time was at a premium. So, early in the twentieth century, the book method was upgraded to point and figure, and the chart looked like Figure 5.1B.

Due to its accuracy in providing trading signals, interest in this unusual type of chart has been continuously growing. Futures traders have used point and figure charting on an intraday basis consistently along with the bar charts. But it was the electronic charting of the 1980s that brought this system into prominence.

The point and figure chart simultaneously emphasizes significant price changes and dismisses low volatility. As discussed in Chapter 1, a 1 × 3 point and figure chart hides insignificant price fluctuations, up to 38 pips, in order to filter out potentially wrong signals. By the same

123

120.00		*120.00*		*120.00*	
119.90	*119.90*	*119.90*	*119.90*	*119.90*	*119.90*
119.80	*119.80*	*119.80*	*119.80*	*119.80*	*119.80*
119.70	*119.70*	*119.70*	*119.70*		*119.70*
119.60	*119.60*	*119.60*	*119.60*		*119.60*
119.50	*119.50*	*119.50*			*119.50*
119.40	*119.40*				*119.40*
119.30					*119.30*
119.20					*119.20*
119.10					*119.10*
119.00					*119.00*
					118.90
					118.80

A - *Late 19th century*

X		X		X		120.00
X	O	X	O	X	O	119.90
X	O	X	O	X	O	119.80
X	O	X	O		O	119.70
X	O	X	O		O	119.60
X	O	X			O	119.50
X	O				O	119.40
X					O	119.30
X					O	119.20
X					O	119.10
X					O	119.00
					O	118.90
					O	118.80

B- Early 20th century

Figure 5.1. The original book method (A) in the late nineteenth century and the same entries marked in a typical point and figure chart (B) in the early twentieth century.

token, it generates clear trading signals that are unadulterated by extended, lethargic, and slow moving prices. Whereas all other charts plot the price in terms of time periods, the point and figure chart plots only the price activity. Time is irrelevant for the purposes of this chart. Some chartists mark the time periods, and some electronic financial charting provides the time periods. Since each day has a different number of entries, the width of each day on the chart fluctuates greatly. During "flat" markets there may be very few or no entries for the standard periods of time used in other types of charts. This can be a difficult concept: We might get used to flatter lines or short bars, but not to no chart entry at all. That is why the point and figure chart is a purer approach to charting. In addition, just like the other continuous chart, the line chart, it reveals more price fluctuations disguised by the bar chart.

The point and figure chart was therefore designed to plot price activity better than other types of charts. Toward that end, one of the important things this chart does is to *avoid plotting the lack of activity.* Irrelevant market activity, the so-called *market noise,* is filtered out. To understand how it works, remember how the chart is plotted. The rising prices are marked with Xs and the falling prices are entered as Os, as in Figure 5.2.

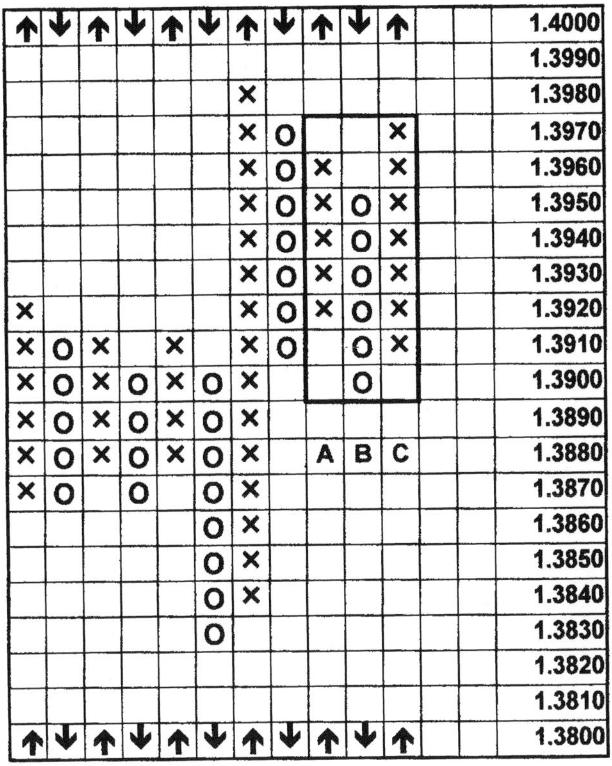

													Price
↑	↓	↑	↓	↑	↓	↑	↓	↑	↓	↑			1.4000
													1.3990
						X							1.3980
						X	O			X			1.3970
						X	O	X		X			1.3960
						X	O	X	O	X			1.3950
						X	O	X	O	X			1.3940
						X	O	X	O	X			1.3930
X						X	O	X	O	X			1.3920
X	O	X		X		X	O		O	X			1.3910
X	O	X	O	X	O	X	O		O				1.3900
X	O	X	O	X	O	X							1.3890
X	O	X	O	X	O	X		A	B	C			1.3880
X	O		O		O	X							1.3870
					O	X							1.3860
					O	X							1.3850
					O	X							1.3840
					O								1.3830
													1.3820
													1.3810
↑	↓	↑	↓	↑	↓	↑	↓	↑	↓	↑			1.3800

Figure 5.2. A typical point and figure chart.

Let's take the very popular example of a *1 × 3 point and figure chart* (also known as the *3-box reversal*). Every price unit that continues the original direction is entered. When the market changes direction, no entry is posted before the price reverses by 3 price units. In currency trading a price unit is generally 10 pips. At that point, all 3 price units are recorded, along with all the additional price units that maintain the direction, on an individual basis. This way, all the insignificant moves are tuned out, allowing the trader to focus on the real price behavior.

In the framed segment of Figure 5.2, the established direction is up (Xs). In column A, the price range is between 1.3920–1.3960 and 1.3920–1.3969. Every time the price advances another 10 pips, such as 1.3940 or 1.3950 or 1.3960, you mark it with one X. The direction turns in column B (Os) only when 1.3930 was traded. Any price activity between 1.3931 and 1.3969 is disregarded. Only at the moment when 1.3930 is traded, you or your system enters three consecutive Os: 1.3950,

1.3940, and 1.3930. After this triple entry, every time the price falls another 10 pips, such as 1.3920 or 1.3910 or 1.3900, you mark it with one O. The new trading range on the down side is between 1.3959 and 1.3891. The direction changes again to up in column C (Xs) when 1.3930 is paid. Once 1.3930 is paid, you enter three consecutive Xs: 1.3910, 1.3920, and 1.3930. Again, you mark an X when the price reaches a new high on this move: 1.3940, 1.3950, 1.3960, and 1.3970.

In technical analysis in general and for point and figure charting in particular, the number 3 is significant. Previously, empirical evidence pointed to the importance of this number—the head and shoulders or the triple top and triple bottom formations. In the following point and figure chart formations, you will see that the number 3 becomes prominent. There is no known scientific reasoning behind this phenomenon, but it is a reality that has proved itself worth observing. See Figure 5.3.

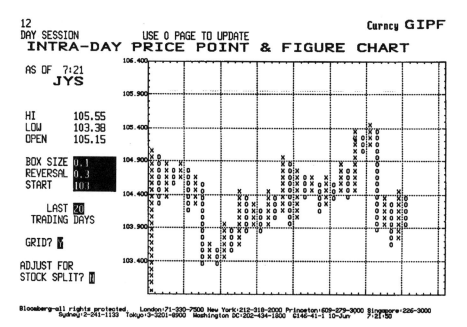

Figure 5.3. A point and figure chart in US dollar/Japanese yen. (*Source:* Bloomberg Financial Services.)

Incidentally, it is possible to use different reversal combinations. A 1 × 1 reversal, for instance, may not be the best possible choice, because the trading signals lose some of their sensitivity due to statistical noise. As a rule of thumb (which should be adjusted for personal preference), the higher the number of reversal boxes, the sharper the sensitivity becomes to the more significant price fluctuations. Moreover, the number of pips per box can differ from 10. In a highly volatile currency, you may want to increase the number of pips per box to 20 in order to have a better view of your chart.

A more common combination outside the standard 1 × 3 is 1 × 2, with the size of the box set at 50 pips. This combination works reasonably well on long-term point and figure charts.

Technologically, compressing the image on the electronic charting services without decreasing the image quality is very difficult. Some services give you the option of replacing the alternating columns of Xs and Os with alternating arrows heading in opposite directions.

Volume is not recorded or available in the forex cash market, as it is in the currency futures market, but it is an integral part of the chart. Since volume creates trading activity, the heavier the volume is, the more activity is charted. Therefore, volume is alive and well, albeit indirectly.

We mentioned in Chapter 1 the importance of the 45-degree trend line. In point and figure charting, trend lines must always be 45-degree lines. The support and resistance lines may occur at different angles.

Point and Figure Charting Formations

Simple Formations

The most basic trading signals in point and figure charting occur when the price makes a new 10-pip high or low. In Figure 5.4A, the original high in USD/JPY is 119.80. The market reverses to 119.50 and then rebounds, advancing through the initial high and establishing a new high at 119.90. This advance to a new high should generate a buying signal. In Figure 5.4B, the opposite is true. The initial direction is down and the current low is 119.60. USD/JPY rebounds temporarily to 119.90 before it continues to fall to a new low of 119.50. Reaching a new low should generate a selling signal.

Unfortunately, the simple formations have a high rate of failure. Whereas making new highs or lows should be significant developments, these patterns may also occur either within the range or at the extreme of the range on very low volume due to overshooting. Either way, in the previous cases the signals lose greatly in significance. The larger players use them typically as bull or bear traps. Consequently, the trader should exert special caution if using solely this type of signal.

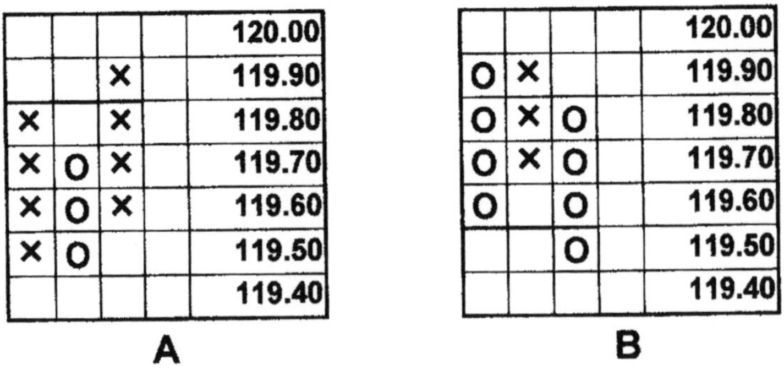

				120.00
	X			119.90
X		X		119.80
X	O	X		119.70
X	O	X		119.60
X	O			119.50
				119.40

A

				120.00
O	X			119.90
O	X	O		119.80
O	X	O		119.70
O		O		119.60
		O		119.50
				119.40

B

Figure 5.4. A typical simple buying signal (A) and a typical simple selling signal (B).

Breakout of a Triple Top

The *breakout of a triple top* formation consists of three consecutive rallies; the first two up side attacks stop at the same price level and the third penetrates the resistance line. Since the original two rallies are capped at the same price, they generate a horizontal resistance line. The breakout from the consolidation range (also known as the *fulcrum*) through the resistance line is followed by a currency rally equal to the size of the bottoms. In Figure 5.5, GBP/USD advances to the high of 1.4980 twice and then

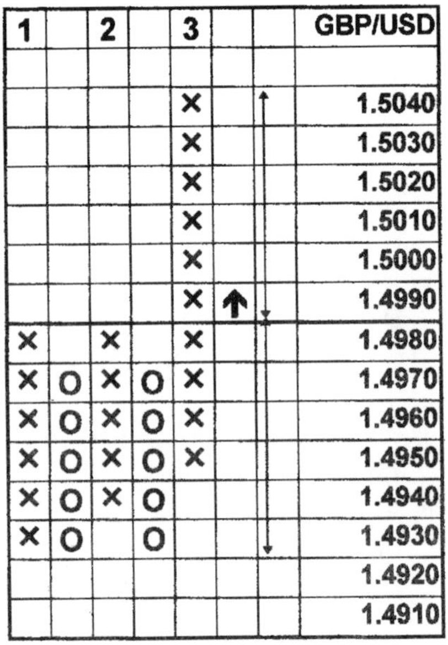

1	2	3			GBP/USD
		X		↑	1.5040
		X			1.5030
		X			1.5020
		X			1.5010
		X			1.5000
		X	↑		1.4990
X		X	X		1.4980
X	O	X	O	X	1.4970
X	O	X	O	X	1.4960
X	O	X	O	X	1.4950
X	O	X	O		1.4940
X	O		O	↓	1.4930
					1.4920
					1.4910

Figure 5.5. A typical breakout of a triple top.

reverses to a low of 1.4930. The third time the currency rallies, it breaks the 1.4980 resistance line on high volume and rallies to a high of 1.5040. The high of 1.5040 represents the price objective of this pattern. To calculate it, add the number of vertical boxes (1.4930 − 1.4980 ⇒ 6 boxes) of the up move below 1.4980 to the neckline at the breakout point (1.4990 − 1.5040 ⇒ 6 boxes, or 60 pips). The trigger point is 1.4990, a price that must attract significant volume in order to validate the breakout. Figure 5.6 displays a market example.

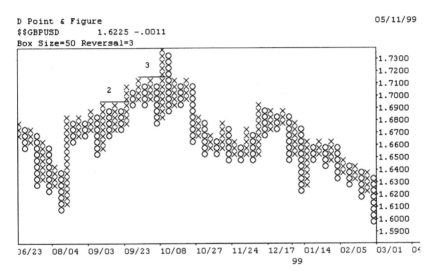

Figure 5.6. Examples of breakout of a triple top formation on the sterling/US dollar chart. (*Source:* Bridge Information Systems, Inc.)

Breakout of a Triple Bottom

The *breakout of a triple bottom* formation consists of a series of three consecutive attacks on the down side: The first two are capped at the same price level, and the third results in the penetration of the support line. The support line is horizontal because the first two dips stop at the same price level. The breakout of this support line on high volume is followed by a currency fall, with a price objective equal in size to the height of the tops.

In Figure 5.7, AUD/USD trades in a range of .7500–.7440 (7 boxes) and reaches the low of .7440 three times. The third time, the market breaks through the .7440 support line on high volume and reaches its price objective at the low of .7370, which is 7 boxes down from .7440. The drop below .7440 is validated by hitting .7430.

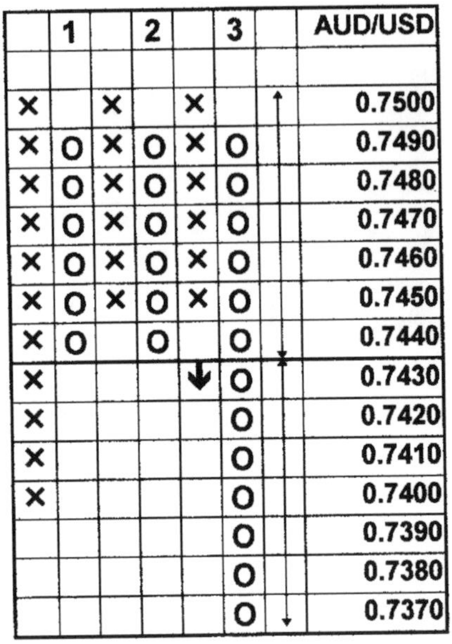

	1		2		3			AUD/USD
X		X		X			↑	0.7500
X	O	X	O	X	O			0.7490
X	O	X	O	X	O			0.7480
X	O	X	O	X	O			0.7470
X	O	X	O	X	O			0.7460
X	O	X	O	X	O			0.7450
X	O		O		O		↓	0.7440
X				↓	O		↑	0.7430
X					O			0.7420
X					O			0.7410
X					O			0.7400
					O			0.7390
					O			0.7380
					O		↓	0.7370

Figure 5.7. A breakout of a triple bottom formation.

Breakout of a Spread Triple Top

The *breakout of a spread triple top* pattern is a spinoff of the breakout of a triple top. The difference is that the third consecutive Xs column fails to reach the horizontal resistance line that was defined by the previous two currency rallies. However, the next (fourth) attack not only reaches the high for the third time, but also breaks through the resistance level if volume is high. The up side target of this breakout from the neckline is approximately the price range of the bottoms, measured from the breakout point.

In Figure 5.8, a USD/JPY move ranged between 100.00 and 100.60 (70 pips) and topped twice at the high. USD/JPY failed in its third rally to advance past 100.30, and it was temporarily sold off. The fourth rally reaches the 100.60 resistance line and breaks through it on high volume. USD/JPY advances 70 pips to its price objective of 101.30.

1		2		X		3		USD/JPY
						X	↑	101.30
						X		101.20
						X		101.10
						X		101.00
						X		100.90
						X		100.80
					↑	X	↓	100.70
X		X				X		100.60
X	O	X	O			X		100.50
X	O	X	O	X		X		100.40
X	O	X	O	X	O	X		100.30
X	O	X	O	X	O	X		100.20
X	O		O	X	O	X		100.10
X			O		O			100.00

Figure 5.8. A typical breakout of a spread triple top formation.

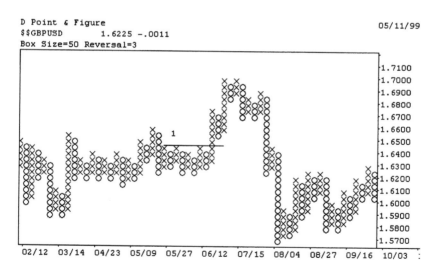

Figure 5.9. Example of breakout of a spread triple top formation on the sterling/US dollar chart. (*Source:* Bridge Information Systems, Inc.)

Figure 5.9 shows an example of a breakout of a spread triple top in sterling/US dollars.

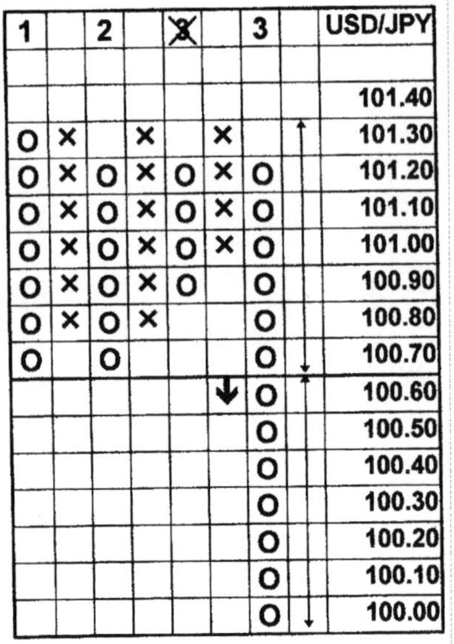

1		2		☒		3		USD/JPY
								101.40
O	×		×		×		↑	101.30
O	×	O	×	O	×	O		101.20
O	×	O	×	O	×	O		101.10
O	×	O	×	O	×	O		101.00
O	×	O	×	O		O		100.90
O	×	O	×			O		100.80
O		O				O	↓	100.70
					↓	O	↑	100.60
						O		100.50
						O		100.40
						O		100.30
						O		100.20
						O		100.10
						O	↓	100.00

Figure 5.10. A typical breakout of a
spread triple bottom formation.

Breakout of a Spread Triple Bottom

The *spread triple bottom* formation is derived from the breakout of a triple bottom formation. A currency fails to fall below a horizontal support level during the third consecutive selloff. However, after a temporary recovery, the currency turns bearish again, reaches the original support level for the third time, and continues its fall to a target equal to the height of the previous consolidation. The breakout of the neckline is followed by a currency rally equal to the height of the tops.

In Figure 5.10, USD/JPY ranges between 100.70 and 101.30 (70 pips). USD/JPY fails to fall below 100.70 the third consecutive time. Its recovery is only temporary. On the next selloff, USD/JPY not only reaches 100.70 for the third time, but penetrates it on high volume and drops 70 pips to 100.00, reaching its price objective. The formation is validated when the market trades 100.60 just under the support line. See Figures 5.11 and 5.12 for examples.

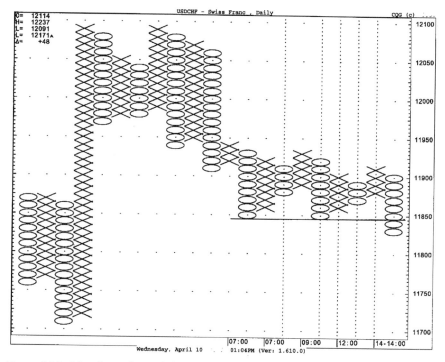

Figure 5.11. A breakout of a triple bottom formation in US dollar/Swiss franc. (*Source:* CQG. ©Copyright CQG INC.)

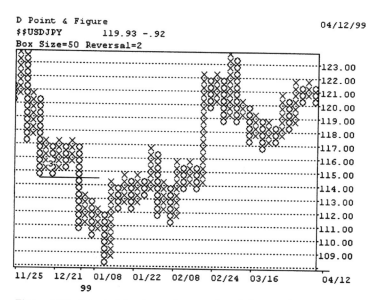

Figure 5.12. A typical breakout of a spread triple bottom formation in US dollar/Japanese yen. (*Source:* Bridge Information Systems, Inc.)

Ascending Triple Tops

The *ascending triple tops* formation is also a spinoff of the breakout of a triple tops formation. In the triple tops formation, the first two rallies stop at the same price level and generate a horizontal resistance line. In the ascending triple tops formation, each new consecutive top is higher than the previous one, thus generating a rising resistance line. The third top is generally a buying signal. The break of the resistance line confirms a buying signal.

In Figure 5.13, the currency gets stronger and prepares to break out as it reaches first 1.4970 in Xs column 1 and then 1.4980 in Xs column 2. Based on the price information at the top of column 2, and the example in Figure 5.4, you might perceive a buying signal. However, volume is not sufficient at this moment to generate a breakout. The third rally manages to attract enough buying interest not only to make a new consecutive high at 1.4990, but also to break above the resistance line; 1.5000 validates the breakout. This pattern does not generally provide an exact price objective. The upward breakout is natural since the market has shown its upward bias in the first two rallies. Therefore the market does not need much convincing to buy on the breakout, generating good volume. See Figure 5.14.

1	2	3			
		X			1.5040
		X			1.5030
		X			1.5020
		X	↗		1.5010
	↑	X			1.5000
		X			1.4990
	X	X			1.4980
X	X	O	X		1.4970
X	O	X	O	X	1.4960
X	O	X	O	X	1.4950
X	O		O	X	1.4940
X			O		1.4930
X					1.4920

Figure 5.13. A typical breakout of an ascending triple top formation.

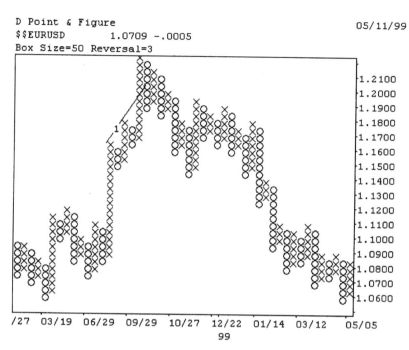

D Point & Figure   05/11/99
$$EURUSD   1.0709 -.0005
Box Size=50 Reversal=3

Figure 5.14. Example of a breakout of an ascending triple top formation on the euro/US dollar chart. (*Source:* Bridge Information Systems, Inc.)

Descending Triple Bottoms

The *descending triple bottoms* formation is another spinoff of the breakout from the triple bottoms formation. Whereas the breakout of the triple bottoms formation was characterized by a horizontal support line, the descending triple bottoms pattern had downward sloping support line because each consecutive bottom is lower than the preceding one. Enough selling interest accumulates in the third Os column to break out through the support line, thus generating a clear selling signal.

In Figure 5.15, a bearish GBP/USD makes two consecutive lower lows at 1.5010 and 1.5000, respectively. The third low not only reaches the next natural low on the support line at 1.4990, but breaks through it on high volume. The descending triple bottoms formation is validated by trading at 1.4980. There is no reliable price objective, but you should loosely target about 60 pips, which is the rough height of the fulcrum. Volume below the support line is good because the direction of the breakout is "in sync" with the previous negative sentiment of the market. See Figure 5.16 for a market example.

	1		2		3		GBP/USD
✗		✗		✗			1.5060
✗	O	✗	O	✗	O		1.5050
✗	O	✗	O	✗	O		1.5040
✗	O	✗	O	✗	O		1.5030
✗	O	✗	O	✗	O		1.5020
✗	O		O	✗	O		1.5010
✗			O		O		1.5000
✗					O		1.4990
✗			↓		O		1.4980
✗					O		1.4970
✗					O		1.4960
✗					O		1.4950
					O		1.4940
					O		1.4930

Figure 5.15. A typical descending triple
bottom formation.

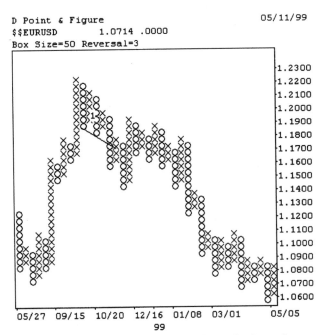

Figure 5.16. Example of a bearish breakout of a descending
triple bottom formation on the euro/US dollar chart.
(*Source:* Bridge Information Systems, Inc.)

Upward Breakout of a Bullish Resistance Line

The *upward breakout of a bullish resistance line* pattern closely resembles the ascending triple top formation, but one or more X columns are added to the chart. In this formation, a currency makes several consecutive highs before it breaks above the bullish resistance line.

The important fact is that the larger number of significant highs enhances the technical importance of this resistance line. The more tangential highs on the line show that a large buying interest is accumulating below it and that is likely to attract further buying power after the breakout and generate a sustained rally.

In Figure 5.17, the buying interest in USD/CAD is accumulating, generating new consecutive highs in columns 1, 2, and 3 before breaking above the bullish resistance line in column 4. Notice that the resistance line and the market interest are "in sync," since they go in the same direction. This combination translates to good market volume following the breakout point. See Figure 5.18.

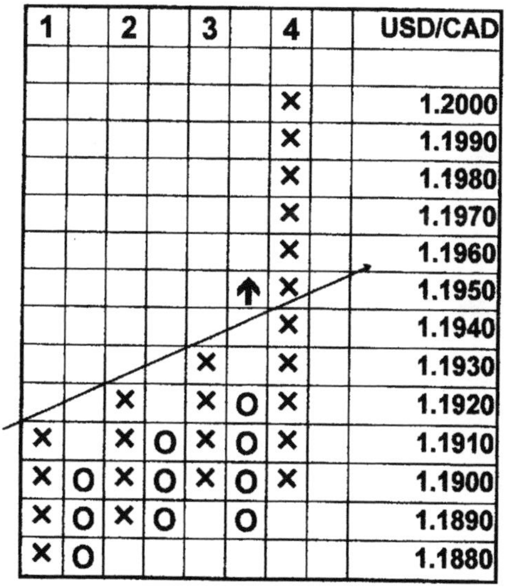

1	2	3	4	USD/CAD
			×	1.2000
			×	1.1990
			×	1.1980
			×	1.1970
			×	1.1960
		↑	×	1.1950
			×	1.1940
		×	×	1.1930
	×	×	O	×	1.1920
×		×	O	×	O	×	1.1910
×	O	×	O	×	O	×	1.1900
×	O	×	O		O		1.1890
×	O					1.1880

Figure 5.17. A typical upward breakout of a bullish resistance line.

Downward Breakout of a Bearish Support Line

The *downward breakout of a bearish support line* is a spinoff of the descending triple bottom formation. The new pattern contains more bearish O columns before the bearish support line breaks. The bearishness of this

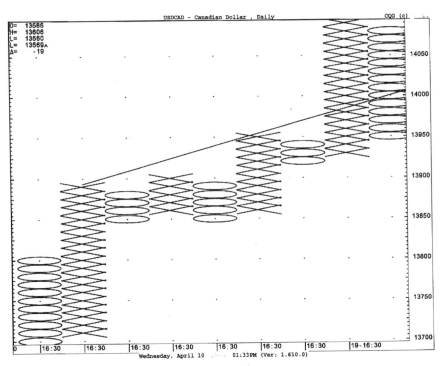

Figure 5.18. An upward breakout of a bullish resistance line in US dollar/Canadian dollar. (*Source:* CQG. ©Copyright CQG INC.)

selling is exacerbated by the accumulation of the negative bias of the market, reflected by the several O columns.

In Figure 5.19, you can see a bearish support line with the lower lows of columns 1, 2, 3 and 4. After a minor retracement, the selling reflected by column 5 accumulates enough selling interest to finally penetrate the bearish support line and generate an extended selloff. See Figure 5.20.

Downward Breakout of a Bullish Support Line

The *downward breakout of a bullish support line* pattern differs from the breakout of a bearish support line formation in the direction of the breakout.

In this pattern, the currency has an upward bias as it moves within a bullish channel. However, the currency surprisingly switches direction due to an unexpected development in the market and performs a valid

1		2		3		4		5		Price
										1.9960
										1.9950
X		X								1.9940
X	O	X	O							1.9930
X	O	X	O	X		X				1.9920
X	O	X	O	X	O	X	O	X		1.9910
X	O		O	X	O	X	O		O	1.9900
X			O		O	X	O	X	O	1.9890
			O		O	X	O		O	1.9880
			O				O		O	1.9870
									O	1.9860
									O	1.9850
									O	1.9840
									O	1.9830
									O	1.9820
									O	1.9810
										1.9800

Figure 5.19. A typical downward breakout of a bearish support line.

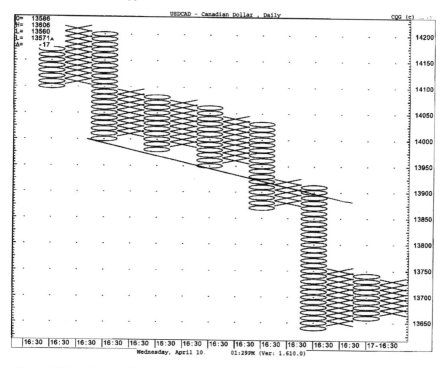

Figure 5.20. A downward breakout of a bearish support line in the US dollar/Canadian dollar. (*Source:* CQG. ©Copyright CQG INC.)

breakout through the support line. The trader faces a bearish signal. The surprise of the move forces traders to start liquidating and reversing their long positions, thus creating a snowball effect in the market.

In Figure 5.21, a seemingly bullish USD/CAD generates several consecutively higher lows in columns 1, 2, 3, 4, and 5 before crashing through the bullish support level.

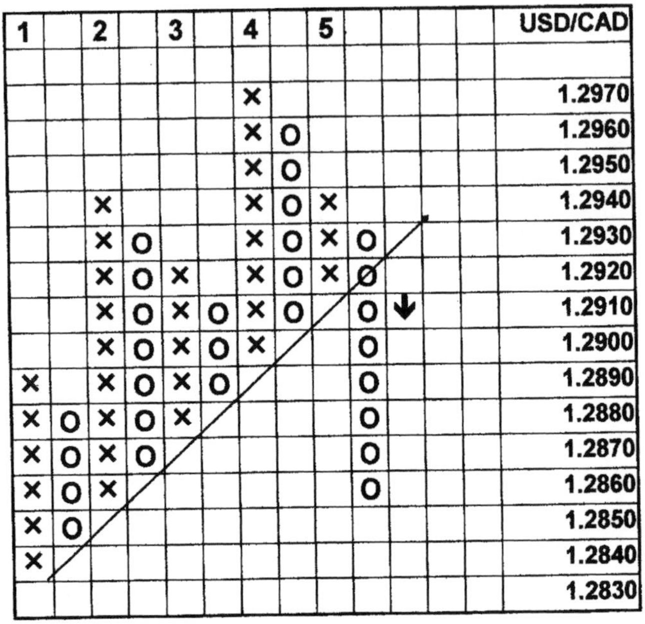

1	2	3	4	5					USD/CAD
			X						1.2970
			X	O					1.2960
			X	O					1.2950
	X		X	O	X				1.2940
	X	O	X	O	X	O			1.2930
	X	O	X	O	X	Ø			1.2920
	X	O	X	O	X	O	O	↓	1.2910
	X	O	X	O	X		O		1.2900
X		X	O	X	O		O		1.2890
X	O	X	O	X			O		1.2880
X	O	X	O				O		1.2870
X	O	X					O		1.2860
X	O								1.2850
X									1.2840
									1.2830

Figure 5.21. A typical downward breakout of a bullish support line.

The bearish breakout is confirmed at 1.2910. Volume tends to be heavy. See Figure 5.22 for a market example.

Upward Breakout of a Bearish Resistance Line

The *upward breakout of a bearish resistance line* pattern is also a surprising formation because it diverges from the preceding market direction. This pattern occurs when the currency breaks out on the up side from the bearish channel. When the breakout is confirmed by heavy volume, this formation generates a buying signal. Naturally, volume tends to be heavy in the beginning because the market has been caught short in a bear trap, and it needs to liquidate and reverse the positions step by step.

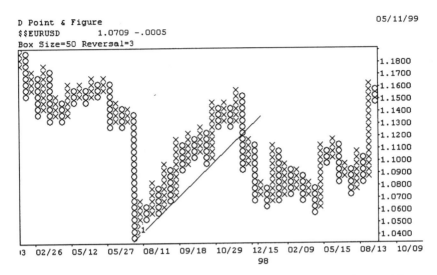

D Point & Figure   05/11/99
$$EURUSD   1.0709 -.0005
Box Size=50 Reversal=3

Figure 5.22. Example of bearish breakout of a rising support line on the euro/US dollar chart. (*Source:* Bridge Information Systems, Inc.)

In Figure 5.23, the USD/JPY mounts a series of consecutively lower highs in columns 1, 2, and 3, reflecting a bearish currency. However, an

1		2		3			USD/JPY
						×	150.00
						×	149.90
						×	149.80
×						×	149.70
×	O					×	149.60
×	O	×				×	149.50
×	O	×	O			×	149.40
×	O	×	O	×		×	149.30
×	O	×	O	×	O	× ↑	149.20
×	O	×	O	×	O	×	149.10
	O		O	×	O	×	149.00
			O		O		148.90
							148.80
							148.70

Figure 5.23. A typical upward breakout through a bearish resistance line.

extraneous factor suddenly changes the psychology of the market and triggers a rally. The upward move is fed by position liquidation and reversal. See Figure 5.24.

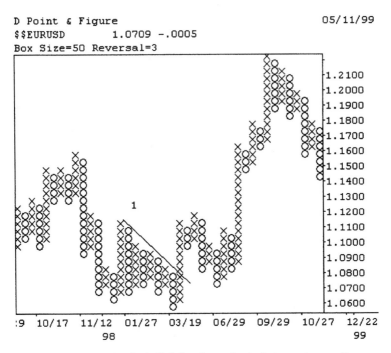

Figure 5.24. Example of a bullish breakout of a declining resistance line on the euro/US dollar chart. (*Source:* Bridge Information Systems, Inc.)

Failures in Point and Figure Charting

You have already seen the special significance of the number 3 in point and figure charting. The third time a price level or a prices series is reached, the market should break in the direction of that price. Conversely, the failure to reach a price or a prices series the third time results in a swift reversal.

As you can see in Figure 5.25, column 3 fails in both cases to reach the previous high of 119.80 (A) and the previous low of 1.4940 (B). Moreover, in Figure 5.25B, columns I, II, and III have the same high of 1.4980. The failure of column III to break on the up side triggers a temporary selloff to 1.4960.

1		2		3		USD/JPY
						119.90
X		X				119.80
X	O	X	O	X		119.70
X	O	X	O	X	O	119.60
X	O	X	O	X	O	119.50
X	O				O	119.40
X					O	119.30
X					O	119.20
X					O	119.10
						119.00
						118.90

A

I		II		III			USD/CAD
						X	1.5010
						X	1.5000
						X	1.4990
X		X		X		X	1.4980
X	O	X	O	X	O	X	1.4970
X	O	X	O	X	O	X	1.4960
X	O	X	O	X	O		1.4950
X	O		O				1.4940
X							1.4930
							1.4920
	1		2		3		1.4910

B

Figure 5.25. Typical failures in point and figure charting.

These failures typically result in the sudden change in the direction of the price. See Figure 5.26.

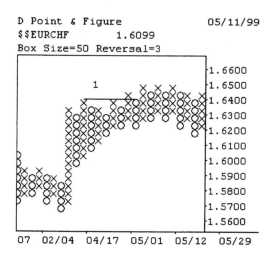

```
D Point & Figure            05/11/99
$$EURCHF          1.6099
Box Size=50 Reversal=3
```

Figure 5.26. Example of a false breakout on the euro/Swiss franc chart. (*Source:* Bridge Information Systems, Inc.)

Common Formations in Point and Figure Charting

The first part of this chapter dealt with the patterns characteristic to the point and figure chart. The second part presents their applications in point and figure charting of trend reversal and trend continuation patterns that are common to all types of charts.

Trend Reversal Formations

Head and Shoulders

In the *head and shoulders* formation, the best-known trend reversal formation, a currency that has swayed above an up trend manages to break the trend line. As a common behavior, the currency retraces toward the trend line, thus creating the illusion of continuing the up trend. The previous trend line, which acted as the most significant support line, turns into a strong resistance line, capping any further advance. There are three peaks—left shoulder, head, and the right shoulder—above a support line, universally known as the neckline (see Chapter 2). The selloff at the end of the third peak penetrates this neckline on high volume and continues the currency reversal to a price objective equal to the height of the middle peak, the head, measured from the breakout point.

Figure 5.27 shows a bearish head and shoulder formation in the GBP/USD. This currency trades in an up trend, generating a series of higher highs. The highest high is in column 2 at 1.7470 (the head). At 1.7400, GBP/USD briefly penetrates the trend line. The following rally, column 3, is weak. It cannot exceed the height of the head, but it approximates the height of column 1 (the left shoulder) at 1.7430. In fact, the rally in column 3 (the right shoulder) is stopped by the previous trend line. The neckline is at 1.7390. The GBP/USD breaks through the neckline and reaches 1.7300, its 90-pip price objective. The head and shoulders reversal is confirmed when 1.7380 is reached on heavy volume. See Figure 5.28.

In the *inverted head and shoulders* formation, the opposite is true. In Figure 5.29, the USD/JPY moves below a downward trend line before briefly penetrating it at 135.00. The low in column 1 (the left shoulder) is 134.60, and the lowest low in column 2 (the head) is 134.30. The USD/JPY attempts to maintain the down trend after breaking out of it. The last selloff in column 3 (the right shoulder) fails to make a new low,

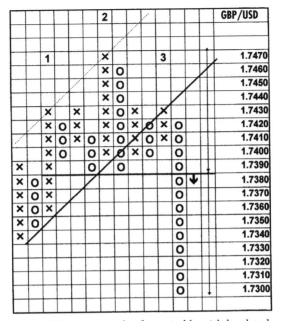

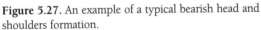

Figure 5.27. An example of a typical bearish head and shoulders formation.

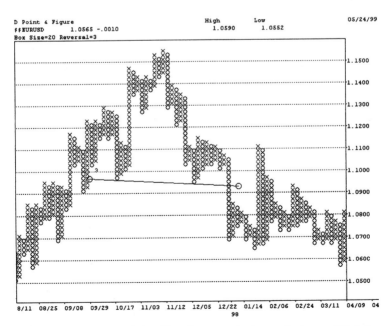

Figure 5.28. A head and shoulders formation in the synthetic euro/US dollar chart. (*Source:* Bridge Information Systems, Inc.)

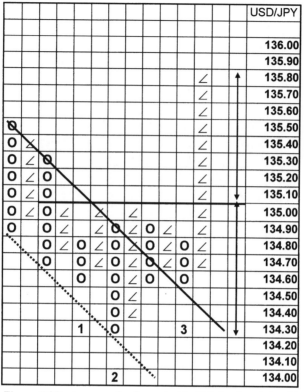

Figure 5.29. An example of a typical bullish head and shoulders formation.

as it is stopped by the previous trend line, now turned into a strong support line. The last selloff is temporary. The neckline is at 135.00. The inverted head and shoulders reversal is confirmed when 135.10 is paid on heavy volume.

Once broken, the USD/JPY rallies to 135.80, its 100-pip price objective, which is equal to the height of the head measured from the breakout point. See Figure 5.30 for a market example.

Triple Top and Triple Bottom

Naturally, the *triple bottom* formation closely mimics the inverted head and shoulders formation. Two differences characterize the triple bottom pattern: (1) All bottoms are equidistant from the neckline, and (2) the price objective is the average length of the bottoms measured from the breakout point. The *triple tops* formation closely resembles the bearish head and shoulders pattern. In the triple tops formation, the tops are of approximately the same magnitude. This characteristic changes the price

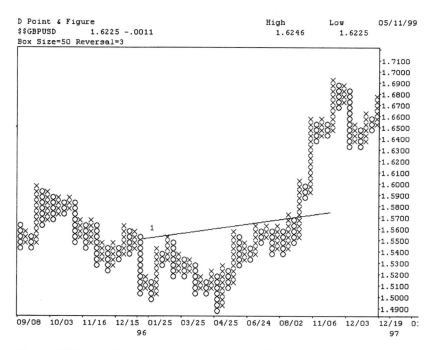

```
D Point & Figure                              High        Low        05/11/99
$$GBPUSD          1.6225 -.0011              1.6246      1.6225
Box Size=50 Reversal=3
```

Figure 5.30. Example of an inverted head and shoulders formation on the sterling/US dollar chart. (*Source:* Bridge Informations Systems, Inc.)

objective to the average height of the three tops measured from the break-out point.

In Figure 5.31, the USD/CAD, which is trading in an up trend, reaches the high of 1.2950 twice above the trend line. After penetrating the trend line, the USD/CAD attempts, in a futile exercise, to continue the up trend. It reaches the high of 1.2950 once more, but it stops in the previous trend line, now turned into a strong resistance line.

The formation has a neckline at 1.2900. The triple top reversal is validated when 1.2890 is reached on heavy volume. The market sells the USD/CAD, and the currency falls 60 pips below the neckline to 1.2840. See Figure 5.32.

In Figure 5.33, the USD/CAD trades within a down trend. After bottoming out twice at 1.2850, the USD/CAD rebounds above the trend line. A last attempt to continue the down trend is stopped by the previous trend line at 1.2850. Thus generating a triple bottom, USD/CAD rebounds through the neckline at 1.2900 on high volume, reaching its

	1				2			3					USD/CAD
	∠			∠			∠						
	∠	O	∠		∠	O		∠	O				1.2950
∠		∠	O	∠	O	∠	O	∠	O				1.2940
∠		∠	O	∠	O	∠	O		∠	O			1.2930
∠		∠	O		O	∠	O	∠	O	∠	O		1.2920
∠		∠	O		O	∠	O	∠	O				1.2910
∠		∠			O		O	O		O			1.2900
∠		∠							O				1.2890
∠	O	∠							O				1.2880
∠	O	∠							O				1.2870
∠	O								O				1.2860
∠									O				1.2850
									O				1.2840

Figure 5.31. A typical triple top formation.

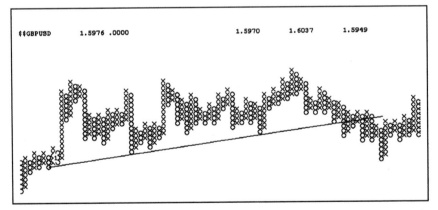

$$GBPUSD   1.5976 .0000   1.5970   1.6037   1.5949

Figure 5.32. Example of triple top formation. (*Source:* Bridge Information Systems, Inc.)

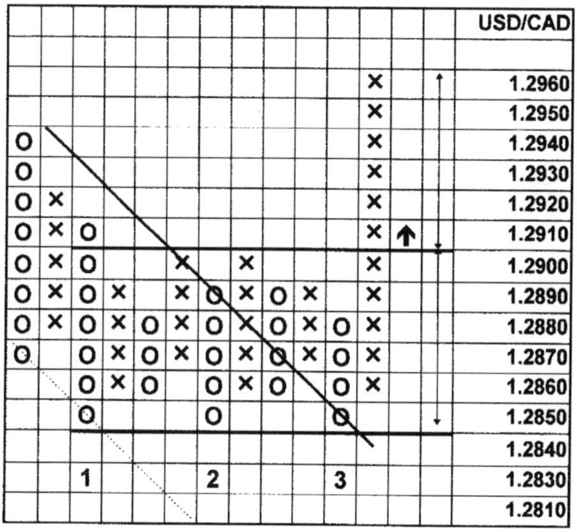

														USD/CAD
										X		↑		1.2960
										X				1.2950
O										X				1.2940
O										X				1.2930
O	X									X				1.2920
O	X	O								X	↑			1.2910
O	X	O			X		X			X				1.2900
O	X	O	X		X	X	X	O	X		X			1.2890
O	X	O	X	O	X	O	X	O	X	O	X			1.2880
O		O	X	O	X	O	X	O	X	O	X			1.2870
		O	X	O		O	X	O		O	X			1.2860
		O			O			O				↓		1.2850
														1.2840
	1			2				3						1.2830
														1.2810

Figure 5.33. A typical triple bottom formation.

60-pip target at 1.2960. The triple bottom reversal is validated when 1.2910 is paid on heavy volume.

Double Top and Double Bottom

The *double top* formation is simpler than the head and shoulders. A currency trending upward breaks the trend line, rebounds temporarily attempting to continue the broken trend, reaches the previous high under the old trend line, and reverses aggressively through a major support line, also known as the neckline. The currency falls under the neckline to a price range equal to the average height of the two tops from the neckline. As usual, the price target is measured from the breakout point.

In Figure 5.34, the GBP/USD trades in an up trend. After topping out at 1.7260, the GBP/USD breaks out of the trend at 1.7200. The market attempts to take the currency back into the up trend, reaching 1.7260 once more, but the previous trend line stops any further advance and the currency turns bearish. The trend is broken. The neckline, which is at 1.7100, is penetrated on heavy volume, and the GBP/USD continues to drop, to reach its target at 1.6960. The double top reversal is validated when 1.2880 is given on heavy volume.

Figure 5.35 illustrates a double top formation in the euro/sterling point and figure charting.

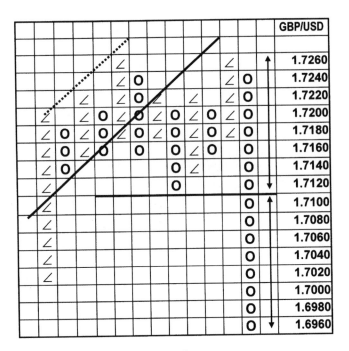

Figure 5.34. A typical double top formation.

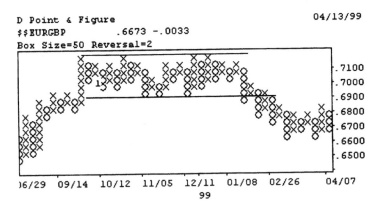

Figure 5.35. A double top formation in the euro/sterling.
(*Source:* Bridge Information Systems, Inc.)

The reverse is true for the *double bottom* formation. A currency trading in a down trend drops to the same low in two different selloffs. However, the second bottom is reached after penetrating the trend line. The down trend is over. The currency breaks out on heavy volume above the neckline, which provided resistance for both bottoms, rallying to a price objective equal to the average size of the bottoms to the neckline.

In Figure 5.36, a down trending USD/JPY bottoms out at 141.40. The currency rebounds above the trend line but the market sells it back. The USD/JPY falls once more to 141.40, but the previous trend line provides significant support. The market covers its short positions and gets long, helping the USD/JPY to penetrate the neckline at 142.00 and advance 70 pips further to reach its 142.70 objective. The double bottom reversal is validated when 142.10 is paid on heavy volume.

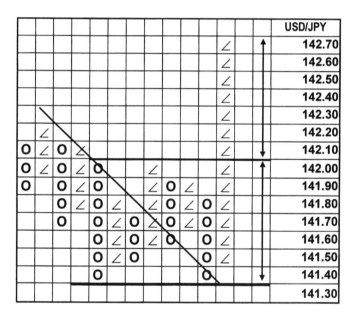

Figure 5.36. A typical double bottom formation.

A market example of a double bottom is provided in Figure 5.37 in the Australian dollar/US dollar.

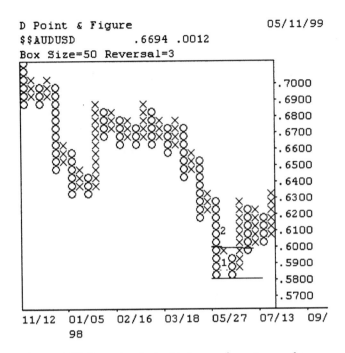

```
D Point & Figure                    05/11/99
$$AUDUSD          .6694 .0012
Box Size=50 Reversal=3
```

Figure 5.37. Example of a double bottom formation on the Australian dollar/US dollar chart. (*Source:* Bridge Information Systems, Inc.)

Rectangle Reversals

A *rectangle reversal* formation has a fulcrum (a distinct congestion formation) that, not surprisingly, resembles a rectangle. In a bullish formation, the resistance line must be penetrated on heavy volume, and the market rallies to a high equal to the range of the fulcrum. In a bearish formation, the support line must be penetrated on heavy volume, and the market falls to a low equal to the range of the fulcrum. The targets are measured from the breakout points.

Figure 5.38 depicts a *bullish rectangle reversal* formation. EUR/USD consolidates in a 40-pip range between 1.1380 and 1.1410 at the end of a move on the down side. When the resistance line of 1.1410 is penetrated on high volume, the EUR/USD rallies for 40 pips to its target of 1.1450. The bullish rectangle reversal is validated when 1.1420 is paid on heavy volume.

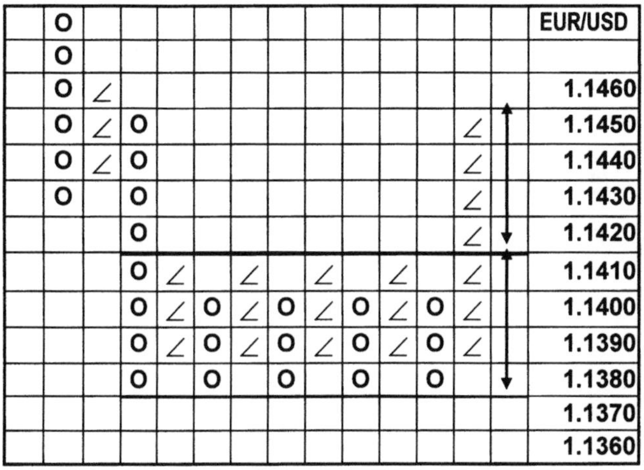

											EUR/USD
O											
O											
O	∠										1.1460
O	∠	O								∠ ↑	1.1450
O	∠	O								∠	1.1440
O	∠	O								∠	1.1430
O		O								∠ ↓	1.1420
		O	∠	∠	∠	∠	∠		∠ ↑		1.1410
		O	∠	O	∠	O	∠	O	∠	O ∠	1.1400
		O	∠	O	∠	O	∠	O	∠	O ∠	1.1390
			O		O		O		O	O ↓	1.1380
											1.1370
											1.1360

Figure 5.38. A typical bullish rectangle reversal formation.

Figure 5.39 shows a market example of a bullish rectangle reversal in the US dollar/Swiss franc.

Figure 5.40 shows a *bearish rectangle reversal* formation. After consolidating in a 40-pip range between 0.7450 and 0.7480, the market sells the AUD/USD aggressively below its 0.7450 support and reaches its target 40 pips lower at 0.7410. The bearish rectangle reversal is validated when 0.7420 is sold on heavy volume.

You can see a practical example of the bearish rectangle in US dollar/Japanese yen in Figure 5.41.

Trend Continuation Formations

Flag

Flag patterns are popular trend continuation formations. A trend is temporarily put on hold for a period of consolidation. The breakout from the consolidation area in the direction of the previous trend signals that the trend is in excellent shape and that it is likely to extend for a range equal to its original stretch.

In Figure 5.42A, a bullish USD/JPY rises from 88.10 and stops its advance at 89.10 (the flagpole). After consolidating against the trend, the USD/JPY breaks out aggressively at 88.70 and resumes its trend by rallying to 89.70, its target equal to the flagpole, as measured from the breakout point.

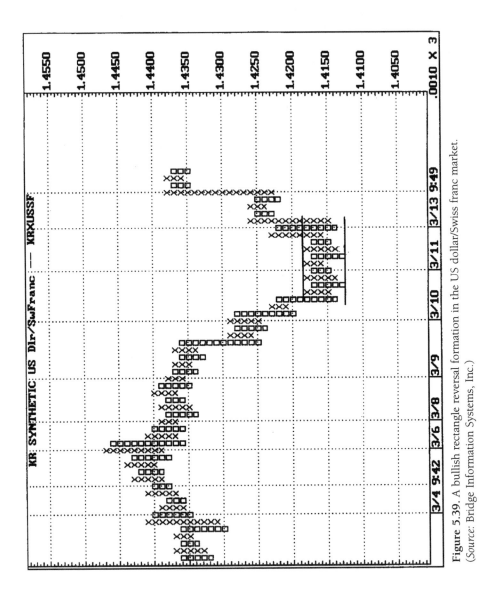

Figure 5.39. A bullish rectangle reversal formation in the US dollar/Swiss franc market. (*Source:* Bridge Information Systems, Inc.)

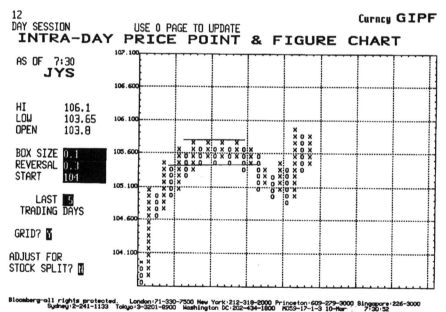

										AUD/USD
			×		×		×		×	↑	0.7480
			×	O	×	O	×	O	×	O	0.7470
			×	O	×	O	×	O	×	O	0.7460
		×		×	O		O		O		O	0.7450
	×		×	O	×					↓	O	0.7440
×	O	×	O	×						O	0.7430
×	O	×	O							O	0.7420
	O									O	↓	0.7410
											0.7400
											0.7390
											0.7380
											0.7370

Figure 5.40. A typical bearish rectangle reversal formation.

Figure 5.41. A bearish rectangle reversal formation in the US dollar/Japanese yen market. (*Source:* Bloomberg Financial Services.)

In Figure 5.42B, a bearish USD/JPY falls from 89.80 to 88.80 (the flagpole) before it consolidates lower. The currency penetrates the support level aggressively at 89.20 and falls to its target of 88.20.

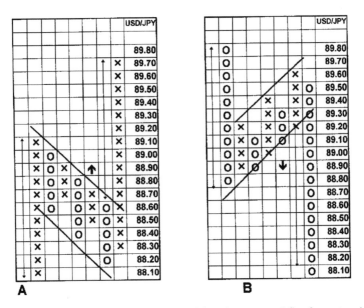

Figure 5.42. A typical flag formation (A) and an inverted flag formation (B).

Figure 5.43 shows a market example of a bull flag formation in the US dollar/Japanese yen.

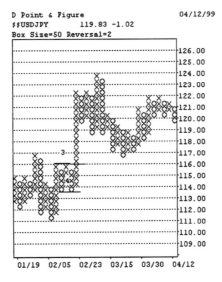

Figure 5.43. A bull flag formation in US dollar/Japanese yen. (*Source:* Bridge Information Services, Inc.)

Pennant

Pennants are closely related to flags. In fact, their only difference is that the consolidation area of a flag resembles a parallelogram, while the pennant looks like, well, a pennant. The target of this formation is the length of the pennant pole measured from the breakout point. For the breakout to be valid, volume must be heavy.

Figure 5.44A presents an example of a pennant in the USD/JPY. The up trend, or the pennant pole, is measured from 88.20 to 89.30. The USD/JPY breaks out of the consolidation area at 88.80 and continues the up trend. The currency rallies to reach its target at 89.90.

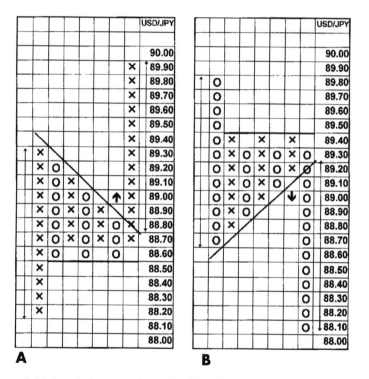

A   **B**

Figure 5.44. A typical pennant formation (A) and an inverted pennant formation (B).

Conversely, Figure 5.44B shows an inverted pennant in the USD/JPY. The down trend, or the pennant pole, is measured from 89.80 to 88.70. The currency breaks out of the consolidation area at 89.20 and resumes its down trend. The USD/JPY falls farther to reach its target at 88.10.

Figure 5.45 illustrates the pennant formation in the US
dollar/Canadian dollar point and figure chart.

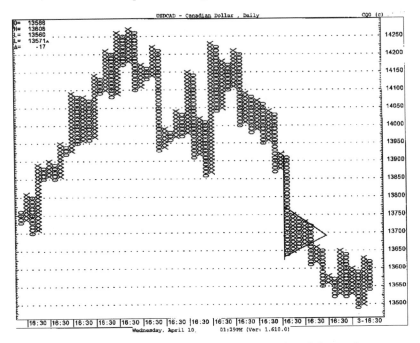

Figure 5.45. A pennant formation in the US dollar/Canadian dollar market.
(*Source:* CQG. ©Copyright CQG INC.)

Breakout of a Symmetrical Triangle

The *breakout of a symmetrical triangle* is a common pattern in point
and figure charting. The symmetrical triangle looks and behaves a lot like
the pennant, except that it lacks the pole. Therefore, the breakout of the
currency—bullish or bearish—has a price objective equal to the base of
the triangle.

Trading this pattern is risky. This chart symmetry denotes that the
market is loosing steam and is looking for a new direction, either way.
There is no previous bias and false breakouts may occur.

In Figure 5.46A, the base of the symmetrical triangle in USD/JPY is
measured between 88.60 and 89.40. This length extrapolates the price
objective to 89.90 as measured from 89.10. Figure 5.46B shows an
inverted symmetrical triangle in USD/JPY. In this case, the base is measured from 89.10 to 89.90. Therefore, the target is 88.60, as measured
from the 89.40 breakout point.

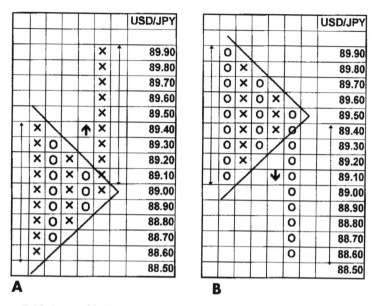

Figure 5.46. A typical bullish symmetrical triangle formation (A) and a typical bearish symmetrical triangle formation (B).

You can see a market example of a symmetrical triangle in the euro/sterling in Figure 5.47.

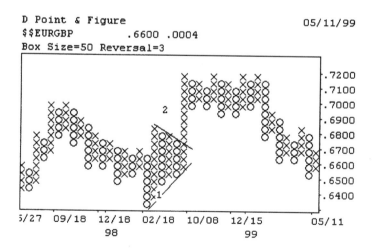

Figure 5.47. Example of symmetrical triangle formation on the euro/sterling chart. (*Source:* Bridge Information Systems, Inc.)

Horizontal Count

The intraday one-box reversal charts offer yet another type of signal, specific only to point and figure charting: the *horizontal count*. The idea behind this approach is that a period of consolidation is generally followed by a sharp move. (Remember: The longer the consolidation period, the sharper is the subsequent price activity.) This method counts the number of columns that form the consolidation area (the horizontal count) and extrapolates, or catapults, this number vertically on top of the breakout price level (see Figure 5.48). The level thus obtained is the new price objective.

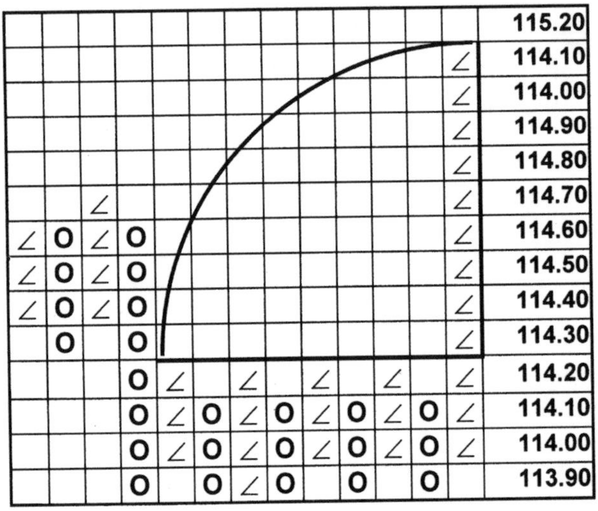

											Price
											115.20
										∠	114.10
										∠	114.00
										∠	114.90
										∠	114.80
		∠								∠	114.70
∠	O	∠	O							∠	114.60
∠	O	∠	O							∠	114.50
∠	O	∠	O							∠	114.40
	O		O							∠	114.30
			O	∠		∠		∠		∠	114.20
			O	∠	O	∠	O	∠	O	∠	114.10
			O	∠	O	∠	O	∠	O	∠	114.00
			O		O	∠	O		O		113.90

Figure 5.48. Diagram of the horizontal count.

Figure 5.49 demonstrates the horizontal count in the euro/Canadian dollar market.

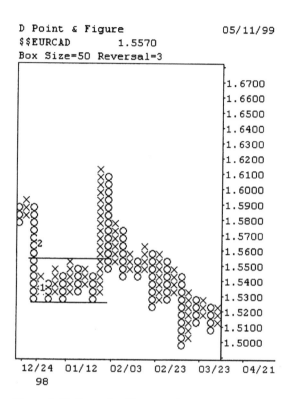

Figure 5.49. Example of horizontal counting on the euro/Canadian dollar chart. (*Source:* Bridge Information Systems, Inc.)

CANDLESTICK CHARTING

One of the most fascinating types of chart is the candlestick chart. Originally developed in China and used in Japan for about three hundred years to forecast the rice prices, candlestick charting is receiving widespread recognition around the world. Its structure is similar to that of the bar chart: Each entry consists of opening, high, low, and closing prices. This chart's advantage is its capacity to provide more trading signals in a user-friendly manner.

The body of the candlestick, or *jittai*, consists of the price range between the opening and closing prices. In their technical interpretation, the Japanese hold that the size and orientation of the candlestick's body provide the most significance regarding the direction of the market. A filled (black) candlestick body indicates that the currency closed lower than it opened. Conversely, a white candlestick body indicates that the currency closed higher than it opened. See Figure 6.1. Originally, the col-

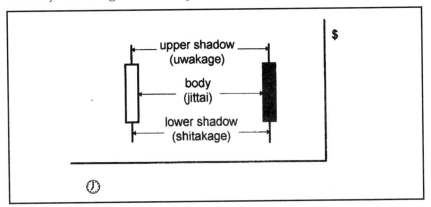

Figure 6.1. The structure of a candlestick.

ors used by the Japanese technicians were red for an up day and black for a down day, and current technology allows for a wide selection of color. However, you only need the up and down days to be marked with different colors. Since directions are differentiated by color, it is easy to quickly grasp the patterns.

The high and low prices create the upper and lower shadows, respectively, when they differ from the open and close prices. The upper shadow, or *uwakage,* occurs when the daily high is higher than either the closing price of a white body or the opening price of a black body. The lower shadow, or *shitakage,* occurs when the daily low is lower than either the closing price of a black body or the opening price of a white body (Figure 6.1).

For example, in Figure 6.2, assume that the USD/JPY opens at 101.00 and closes at 102.00. The high is 102.50 and the low is 100.50. The candlestick's body, determined by the difference between the closing and opening prices, measures 100 pips (102.00 – 101.00). It is white because the closing price (102.00) is higher than the opening price (101.00). The candlestick has an upper shadow between 102.50 and 102.00 and a lower shadow between 101.00 and 100.50.

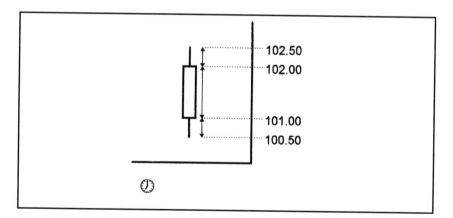

Figure 6.2. The price structure of a candlestick.

The Japanese are fond of nicknames, and candlestick charting has its share. The basic ones are *marubozu, opening bozu,* and *closing bozu.*

A *marubozu,* or *shaved head,* is a candlestick with a full body, but no shadows. This means that in the case of a *white marubozu, the opening price is similar to the low* and *the closing price coincides with the high.* In the case of a *black marubozu, the opening price is similar to the high (yoritsuki takane)* and the *closing price is the low (yasunebike)*.

An *opening bozu* means that, in the case of a white candlestick (an up day), the market opens at the low of the day and therefore there is no lower shadow. In the case of a black candlestick (a down day), the market opens at the high of the day and therefore there is no upper shadow.

A *closing bozu* means that, in the case of a white candlestick (an up day), the market closes at the high of the day and therefore there is no upper shadow. Conversely, in the case of a black candlestick (a down day), the market closes at the low of the day and therefore there is no lower shadow.

Bullish Signals

Daily White Candlesticks

We will focus first on the *white candlesticks (yo-sen)*, which show that the currency closes higher than it opens. To "read" their signals, remember a simple and practical rule: *The longer the body of the candlestick is, the more bullish is the signal.* In Figure 6.3:

- Candlesticks A and B, with the longest bodies, are the most bullish.
- Candlesticks C, D, and E are "regularly" bullish due to their long bodies and long shadow (either upper or lower).
- Candlestick F should be bullish, but you have to be careful because the tiny upper and lower shadows show possible consolidation.

Figure 6.4. displays several market examples of bullish white candlesticks.

Two-Day Bullish Signals

In addition to simple bullish candlesticks, two-day formations generate bullish signals. The names of the patterns and the buying signals are generated by the second day's candlestick. They are:

- *Kirikomi*, or *kirikaeshi* or piercing line.
- Bullish *tasuki*.
- Upside gap *tasuki*.
- Bullish engulfing pattern, or the bullish *tsutsumi*.

NAME OF CANDLESTICK	NICKNAME	SIGNAL
A. Long white candlestick		☝️☝️
B. Long white candlestick (no shadows)	white marubozu	☝️☝️
C. White candlestick lower shadow		☝️
D. Long white candlestick (lower shadow)	white closing bozu	☝️
E. Long white candlestick (upper shadow)	white opening bozu	☝️
F. Short white candlestick		✋☝️

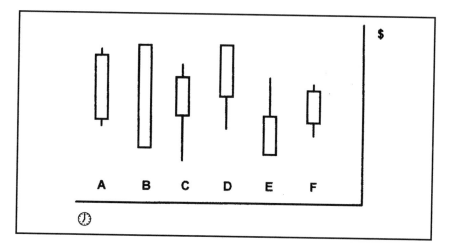

Figure 6.3. Typical bullish white candlesticks.

Kirikomi (Kirikaeshi) Candlestick

The *kirikomi*, or *kirikaeshi*, *candlestick* is part of a two-day candlestick combination that is typical of the currency futures market. A first-day black marubozu candlestick is followed by the kirikomi candlestick—a white marubozu candlestick—which *opens lower than the previous low and closes above the 50% level but below the opening price of the previous day's range* (see Figure 6.5). This pattern is also know as the piercing line.

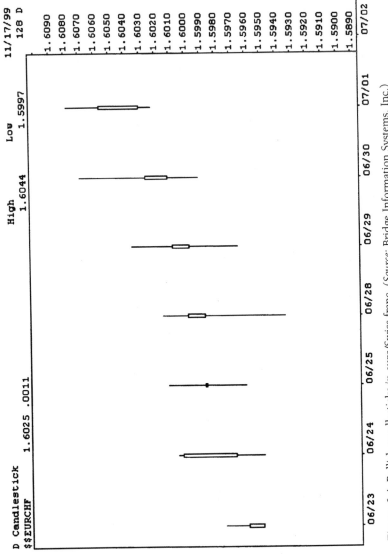

Figure 6.4. Bullish candlesticks in euro/Swiss franc. (*Source:* Bridge Information Systems, Inc.)

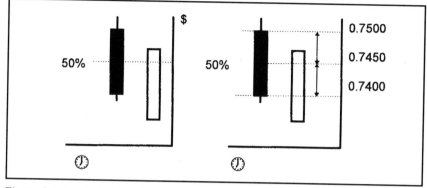

Figure 6.5. A typical *kirikomi* candlestick.

The first range indicates a down day. On the second day, the currency opens lower than the previous day's low, to generate an overnight price gap. Despite the weak opening, the currency does not weaken any further or, if it does, the movement is marginal. The market fills this gap and continues to rally, recovering most of the lost ground from the previous session. In fact, the point size of the two ranges tends to be about the same. By closing above the 50% level of the first day's range, the *kirikomi* candlestick generates a strong buying signal, which is further confirmed when the price of the following session climbs above the highs of the original two days. If the kirikomi candlestick closes above the first day's opening price, it becomes a bullish tsutsumi candlestick, a pattern we will analyze shortly.

Figure 6.6 shows a market example of the *kirikomi* candlestick.

Bullish Tasuki Candlestick

The *bullish tasuki candlestick* is the second leg of a two-day candlestick combination typical of the currency futures market. It consists of *a long black candlestick that opens within the range of the previous day's long white body and closes marginally below the previous day's low* (see Figure 6.7). Candlesticks do not have to have long bodies if the two days' ranges are about the same size.

On the second day of the combination, the market opens lower than the previous close and proceeds to sell the currency further. This down day occurs in an otherwise advancing market and manages to sur-

Figure 6.6. A typical piercing line (kirikomi) in US dollar/Canadian dollar. (*Source:* Bridge Information Systems, Inc.)

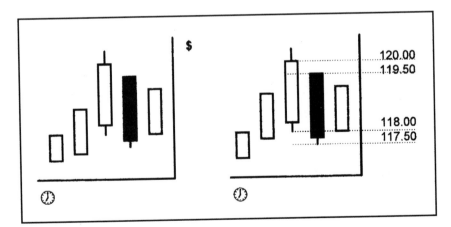

Figure 6.7. A typical bullish *tasuki* candlestick.

prise some of the players. Consequently, the move triggers some profit taking. This is exactly the point of this formation: temporary profit taking in an up trend. The signal of a bullish tasuki candlestick is to continue to buy.

Figure 6.8 shows a market example of the bullish *tasuki* candlestick.

Upside Gap Tasuki Candlestick

The *upside gap tasuki candlestick* occurs in an up trend in the currency futures market. It is *a second-day black candlestick that closes an overnight gap opened on the previous day by a white candlestick*. See Figure 6.9. This pattern is similar to a common gap in the currency futures candlestick chart, which is indeed closed. It provides the trader with a very short-term opportunity to sell to fill the gap, but has no other technical significance. The filling of the upside gap tasuki candlestick signals that *the bullish price activity will resume*.

Figure 6.10 shows a market example of the bullish *tasuki* candlestick.

Bullish Engulfing Pattern (Bullish Tsutsumi)

The *bullish engulfing pattern*, or the *bullish tsutsumi candlestick, is a second-day long white candlestick that opens lower and closes higher than the preceding small black body*. In Figure 6.11, the second day's white candlestick "engulfs" the previous day's range; hence the name of the pattern. This formation is the equivalent of a bullish key reversal in the candlestick charts. Both the *tsutsumi* and the key reversal formations occur in the currency futures markets. The signal is *to reverse to buying*. The *tsutsumi* candlestick can occur in the *spot currency market if the bullish engulfing pattern opens at the low of the previous day's range*. The bullish *tsutsumi* candlestick reflects a general change in the market sentiment and volume tends to be high. Traders must be aware that it is common for a neutral period to immediately follow this formation, because the market needs to digest the large one-day activity and reevaluate the new high price levels.

Bullish *tsutsumi* candlesticks can be observed on the daily euro/Canadian dollar chart in Figure 6.12.

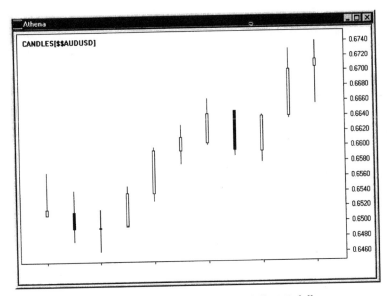

Figure 6.8. A bullish *tasuki* pattern in Australian dollar/US dollar.
(*Source:* Bridge Information Systems, Inc.)

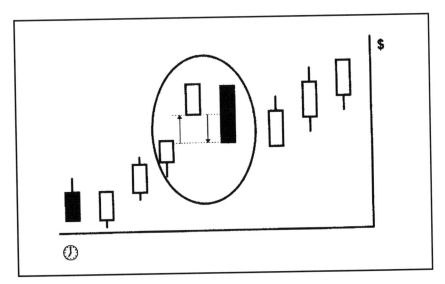

Figure 6.9. The upside gap *tasuki* candlestick.

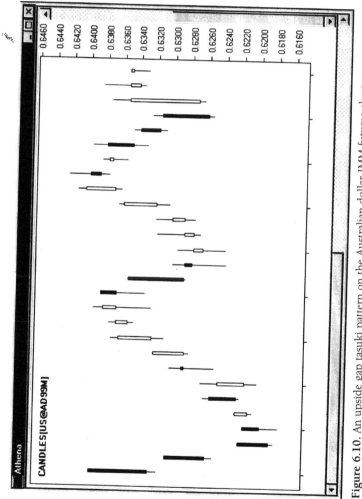

Figure 6.10. An upside gap tasuki pattern on the Australian dollar IMM futures chart. (*Source:* Bridge Information Systems, Inc.)

171

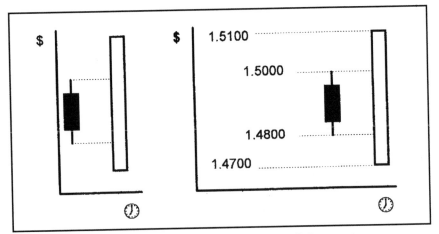

Figure 6.11. The typical bullish engulfing pattern or the bullish *tsutsumi*.

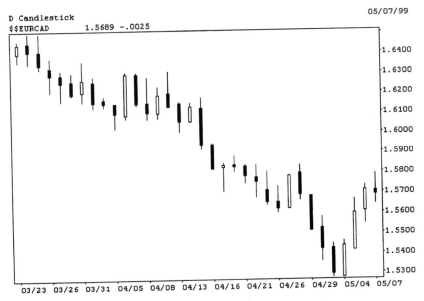

Figure 6.12. A bullish engulfing pattern in the euro/Canadian dollar.
(*Source:* Bridge Information Systems, Inc.)

Bearish Signals

Daily Black Candlesticks

The black or filled (*in-sen*) candlesticks indicate that *the closing price is lower than the opening price.* In Figure 6.13, the longer the body (see A and

B), the more bearish the signal. A long black body with small or no shadows is the weakest. A medium-sized black body, along with a long upper or lower shadow, shows weakness. A medium-sized black body, with tiny upper and lower shadows, retains the weakness bias, but also invites caution since the signal may become murky.

NAME	NICKNAME	SIGNAL
A. Long black candlestick		👇👇
B. Long black candlestick (no shadow)	**Black marubozu**	👇👇
C. Black upper shadow candlestick		👇
D. Long black candlestick (lower shadow)	**Black opening bozu**	👇
E. Long black candlestick (upper shadow)	**Black closing bozu**	👇
F. Bearish short black candlestick		✋👇

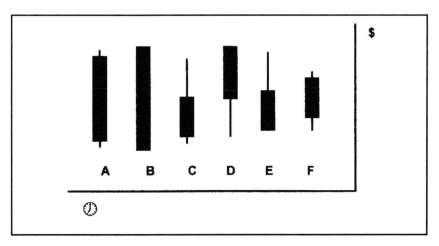

Figure 6.13. Typical black candlesticks.

Several black candlestick market examples are shown in Figure 6.14 in the US dollar/Swiss franc weekly chart.

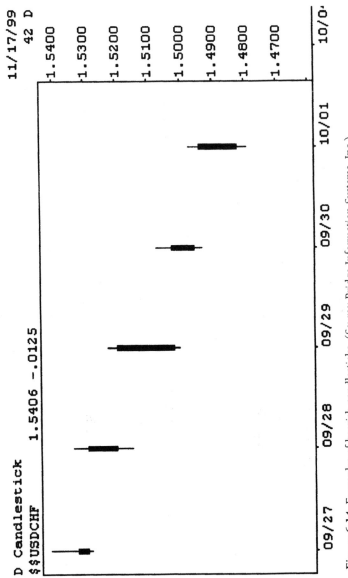

Figure 6.14. Examples of bearish candlesticks. (*Source:* Bridge Information Systems, Inc.)

Two-Day Bearish Signals

In addition to simple bearish candlesticks, two-day formations generate bullish signals. The name of the patterns and the selling signals are generated by the second day's candlestick. They are:

- *Dark cloud cover,* or *kabuse.*
- *Atekubi.*
- *Irikubi.*
- *Sashikomi.*
- Bearish engulfing pattern, or the bearish *tsutsumi.*
- Bearish *tasuki.*
- Downside gap *tasuki.*

Dark Cloud Cover (*Kabuse* Candlestick)

The *dark cloud cover,* or the *kabuse candlestick,* is a two-day pattern characteristic of the currency futures market. An up first day is followed by *a second-day black candlestick that opens at a higher price, thus generating a gap, but that fills the gap and closes midway through the previous day's white long closing bozu body.* The second-day candlestick that negates the original buying activity is called *dark cloud cover,* or *kabuse,* candlestick. The *kabuse* candlesticks generate a *bearish reversal,* or a *selling signal.*

To make it easy to remember, think of this two-day pattern as a one-day combined candlestick in which the original white body was minimized by the selling of the second day. In Figure 6.15, the lower shadow

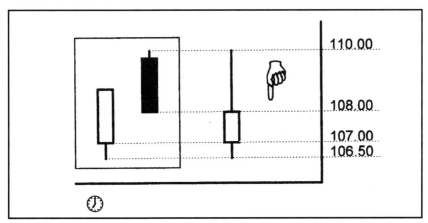

Figure 6.15. Example of the typical dark cloud cover, or the *kabuse* candlestick.

retains the original size, whereas the upper shadow is very long because it equals the entire range of the *kabuse* (the second-day) candlestick.

Based on price behavior on the second day, there are *three types of dark cloud cover patterns* (see Figure 6.16):

- Long black closing *bozu,* which settles under the 50% level of the original white candlestick
- Black *marubozu*
- Long black opening *bozu,* which settles under the 50% level of the original white candlestick

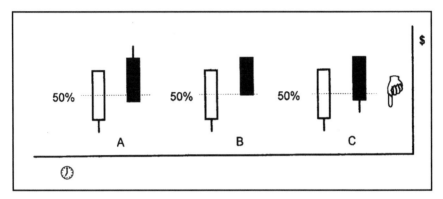

Figure 6.16. Typical dark cloud cover patterns, or *kabuse* candlesticks.

A market example of a dark cloud cover can be observed in Figure 6.17, in the euro/British pound daily chart.

Atekubi (Ate) **Candlestick**

The *atekubi,* or *ate, candlestick* in Figure 6.18 is a second-day small white candlestick that opens lower than the previous day's low, fills the overnight gap, and closes at the daily high. The buying interest is low, and the current closing price at the high of the day equals only the previous day's low. The original day's range is a long black candlestick, and it is generally part of a down trend or a short-term bearish market. This is a typical case of the market closing a common gap in the currency futures market, with no impact on the previous bearish trend. *The completion of the atekubi candlestick signals that the selling will continue.*

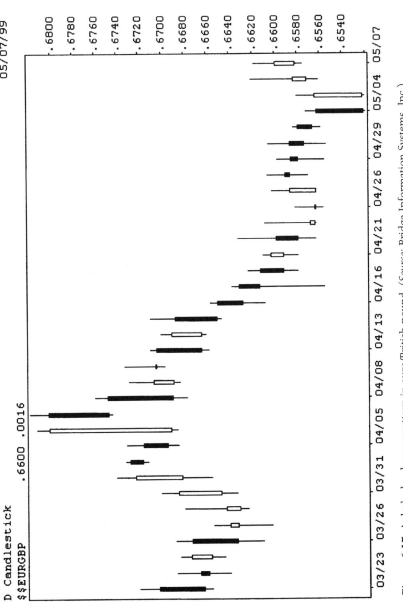

Figure 6.17. A dark cloud cover pattern in euro/British pound. (*Source:* Bridge Information Systems, Inc.)

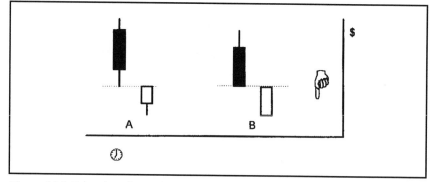

Figure 6.18. Examples of typical *atekubi* candlesticks.

You can see a practical example of an *atekubi* candlestick in Figure 6.19 on the IMM Swiss franc daily futures.

Irikubi Candlestick

The *irikubi candlestick* is a modified atekubi candlestick. All the characteristics are the same, except that the second day's closing high is marginally higher than the original day's low. In Figure 6.20, the *irikubi* candlestick is a second-day short white candlestick that opens lower than the previous day's low, fills the overnight gap, and closes at the daily high. The current closing/high price is marginally higher than the previous day's low of a long black candlestick. This is a typical case of the market closing a common gap in the currency futures market, with no impact on the previous down trend. *The completion of the irikubi candlestick signals further selling.*

Both the atekubi and irikubi candlesticks tend to occur with a higher frequency soon after a reversal or during established bearish markets.

Figure 6.21 show you a market example of an *irikubi* candlestick in the IMM euro futures.

Sashikomi Candlestick

The *sashikomi candlestick* is a modified *irikubi* candlestick. In Figure 6.22, *a second-day white candlestick that opens lower than the previous day's low, fills the overnight common gap and closes at the daily high.* Despite the wider gap thus formed, the white candlestick weakly closes below the 50% range of

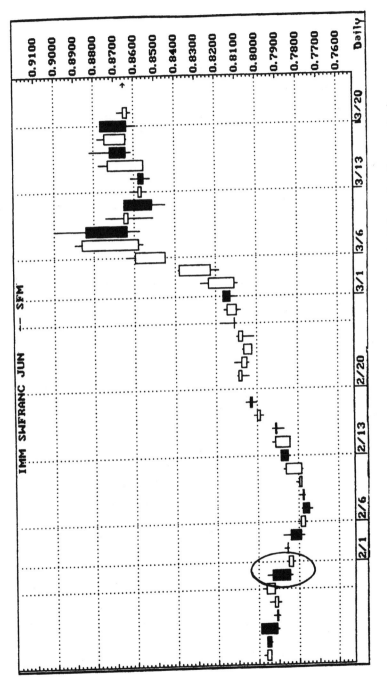

Figure 6.19. *Atekubi* candlestick in the IMM Swiss franc daily futures. (*Source:* Bridge Information Systems, Inc.)

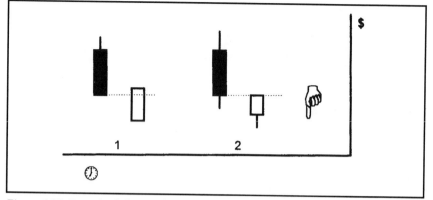

Figure 6.20. Typical *irikubi* candlesticks.

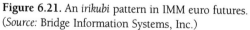

Figure 6.21. An *irikubi* pattern in IMM euro futures.
(*Source:* Bridge Information Systems, Inc.)

the previous day's low. The difference is that *the opening of the sashikomi candlestick is much lower and thus its body is longer than that of the irikubi candlestick.* This price movement is temporary and the trading signal remains *bearish.*

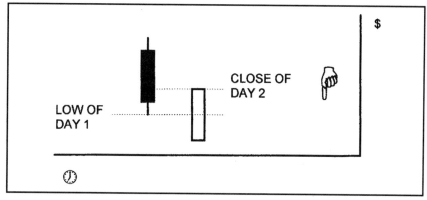

Figure 6.22. Example of a typical *sashikomi* candlestick.

Figure 6.23 shows a market example of a *sashikomi* candlestick in the Canadian dollar IMM futures.

Bearish Engulfing Pattern (Bearish *Tsutsumi*)

The *bearish engulfing pattern*, or the *bearish tsutsumi candlestick*, is *a second-day long black candlestick that opens higher and closes lower than the preceding small white body, thus "engulfing" the previous day's range*. As shown in Figure 6.24, this is the equivalent of a bearish key reversal in the candlestick charts. Both the *tsutsumi* and the key reversal formations occur in the currency futures markets. The signal is *to reverse to selling*. The *tsutsumi* candlestick can occur in the *spot currency market if the bearish engulfing pattern opens at the high of the previous day's range*. The bearish engulfing pattern reflects a general change in the market sentiment and volume tends to be high. Traders may encounter a neutral period immediately following this formation because the market needs to digest the large one-day activity and reevaluate the new low price levels.

A market example of the *bearish engulfing pattern* may be observed in the US dollar/Japanese yen monthly chart in Figure 6.25.

Bearish *Tasuki* Candlestick

The *bearish tasuki candlestick* is the second leg of a two-day candlestick combination. It consists of *a long white candlestick that opens within the body of the previous day's long black body and closes marginally above the previous*

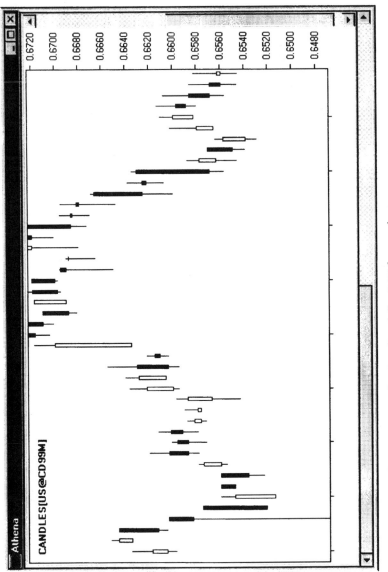

Figure 6.23. A *sashikomi* pattern on the Canadian dollar IMM futures chart. (*Source:* Bridge Information Systems, Inc.)

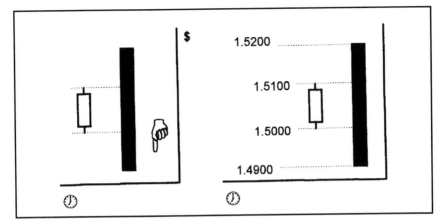

Figure 6.24. A typical bearish engulfing pattern. .

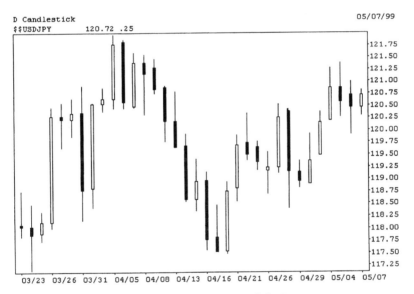

Figure 6.25. The bearish engulfing pattern in US dollar/Japenese yen. (*Source:* Bridge Information Systems, Inc.)

day's low. As shown in Figure 6.26, candlesticks do not have to have long bodies if the two days' ranges are about the same size.

This up day occurs in an otherwise declining market and manages to surprise some of the players. Consequently, the move triggers some profit taking. This is exactly the point of this formation: temporary prof-

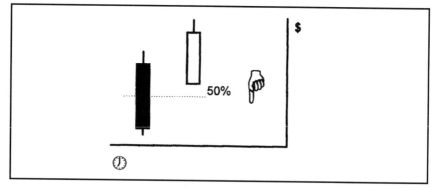

Figure 6.26. A typical bearish *tasuki* candlestick.

it taking in a down trend. The signal of a bullish tasuki candlestick is *to continue to sell.*

Downside Gap *Tasuki* Candlestick

The *downside gap tasuki candlestick* occurs in a down trend. It is *a second-day white candlestick that closes an overnight gap opened on the previous day by a black candlestick* (see Figure 6.27). This is similar to a common gap in the currency futures candlestick chart that is indeed closed. It provides the trader with a very short-term opportunity to buy to fill the gap, but has no other technical significance. The filling of the downside gap *tasuki* candlestick signals that *the bearish price activity will resume.*

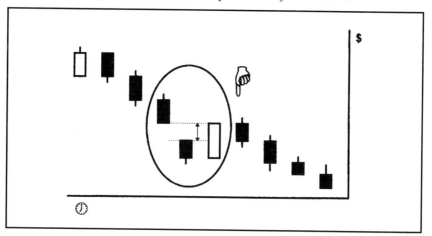

Figure 6.27. A typical downside gap *tasuki* candlestick.

Reversal Signals

Daily Reversal Pattern

The daily reversal patterns in Figure 6.28 are:

- Long-legged shadows' *doji* candlestick
- The hangman and the hammer

NAME	NICKNAME	SIGNAL
A. Opening and closing doji candlestick	**Long-legged shadows' doji**	⤴ or ⤵
B. Opening and closing doji candlestick	**Tonbo (dragonfly)**	⤴ or ⤵
C. Opening and closing doji candlestick	**Tonbo**	⤴ or ⤵
D. Opening and closing doji candlestick	**Tohbu**	⤴ or ⤵ or ✋
E. White lower shadow	**Karakasa** (hangman at the top, hammer at the bottom)	👇 at the top 👆 at the bottom
F. Black lower shadow	**Karakasa** (hangman at the top, hammer at the bottom)	👇 at the top 👆 at the bottom
G. Short white candlestick	**Koma**	👆 or 👇
H. Short black candlestick	**Koma**	👆 or 👇

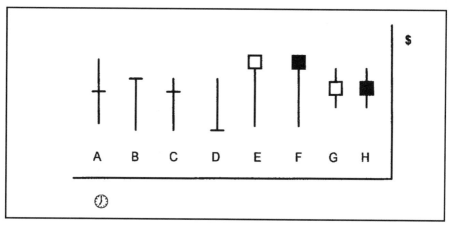

Figure 6.28. Typical daily reversal patterns in candlestick charting.

The nicknames, again, are very interesting. *Komas* are spinning tops, not quite sure which way to go. *Tonbo* is a dragonfly. It signals a reversal, yet it may move either way. *Tohba* or *tohbu* is the gravestone *doji.*

Although these candlesticks may seem to give either bullish signals at the top or bearish signals at the bottom, a closer look and a little caution should help. Remember that *the body is small*—only one-half or one-third of the length of the shadow. Naturally, the trading signals are to sell at the top and buy at the bottom.

Long-Legged Shadows' Doji Candlestick

The *long-legged shadows' doji candlestick* is one of the most recognized candlestick reversal patterns outside Japan and a very reliable reversal formation. A *doji* candlestick resembles a bar. As shown in Figure 6.29, this type of candlestick does not have a body because the opening and closing prices are identical. This shows that the market reaches the end of the trend and a temporary balance between supply and demand before reversing itself. The market tends to reverse immediately after the signal. At times, exceptional pressure on the opposite side may postpone the reversal by one day. Keep in mind that the *doji* candlestick must occur at an extreme level (A) to signal reversal.

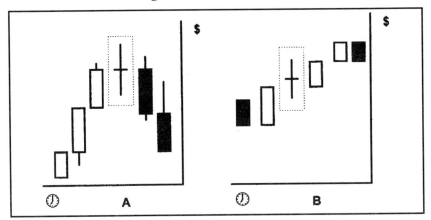

Figure 6.29. A typical *doji* candlestick (A) and a typical rickshaw man (B).

If it occurs in the midrange, the candlestick is called *rickshaw man* (B) and it does not generate any particular trading signal.

Figure 6.30 displays market examples of the *doji* candlesticks in the weekly US dollar/Japanese yen.

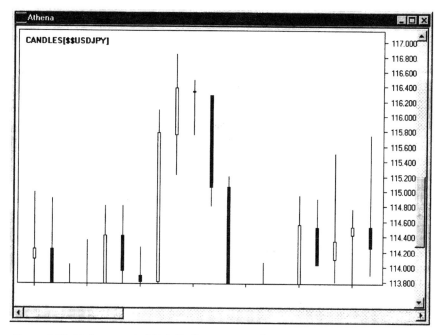

Figure 6.30. Doji pattern in US dollar/Japanese yen chart.
(*Source:* Bridge Information Systems, Inc.)

The Hangman and the Hammer

Initially known as *karakasa,* or paper umbrella, this short candlestick with a very long lower shadow is better known now as the *hangman* when it occurs at the top of the trend, because traders may fall into the trap of buying the currency and being caught long at the top of the range. The opposite is true when this formation occurs at the bottom of the trend. The only difference is the name: the *hammer.* The candlestick can be either white or black.

A market example of a hangman is shown in the British pound/US dollar market in Figure 6.31. You can also see a market example of a hammer on the US dollar/Japanese yen chart in Figure 6.32.

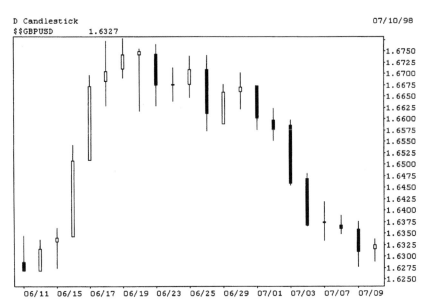

D Candlestick   07/10/98
$$GBPUSD   1.6327

Figure 6.31. A hangman reversal pattern on the British pound/US dollar chart. (*Source:* Bridge Information Systems, Inc.)

Complex Trading Signals

In addition to these several reliable buying, selling, and reversal signals, such as the *doji* candlestick and the engulfing patterns, an amalgam of more sophisticated reversal and continuation patterns are known collectively as *Sakata's 5 methods*. These strategies date back over 200 years. They are:

- *Sanzan* (three mountains)
- *Sansen* (three rivers)
- *Sanpei* (three parallel candlesticks)
- *Sanpo* (three methods)
- *Sanku* (three gaps)

   Note two things:

- The omnipresence of the number 3 (*san* in Japanese). Number 3 has a magnetic attraction among technicians, regardless of geography or time.

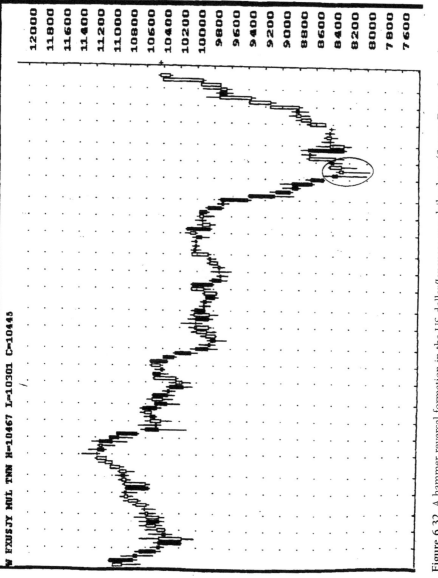

W EXUSJY MUL TNN H=10467 L=10301 C=10445

Figure 6.32. A hammer reversal formation in the US dollar/Japanese yen daily chart. (*Source:* FutureSource.)

- These reversal patterns generally have more daily entries than we were accustomed to so far.

Sanzan (Three Mountains)

Sanzan, or *three mountains,* shown in Figure 6.33, closely resembles the triple top formation. Prices move up in three waves and then descend in three waves. The three tops have relatively the same height. When the support line underneath and adjacent to the three tops, known as the neckline, is penetrated on high volume, the currency reverses from up to down. The price objective is equal to the average height of the tops, measured from the breakout point.

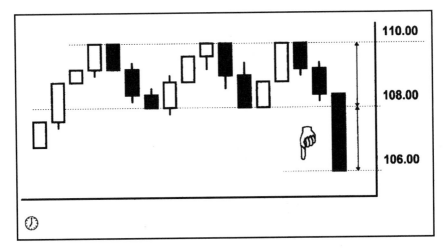

Figure 6.33. A typical three mountains reversal formation.

Figure 6.34 shows a market of the three mountains reversal formation on the British pound/US dollar monthly chart.

If the middle mountain is higher than the other two, the formation is called the three Buddha top formation (see Figure 6.35). This candlestick formation closely resembles the head and shoulders formation. When the neckline is penetrated on high volume, the currency reverses from up to down. Its price objective is equal to the height of the middle top, measured from the breakout point.

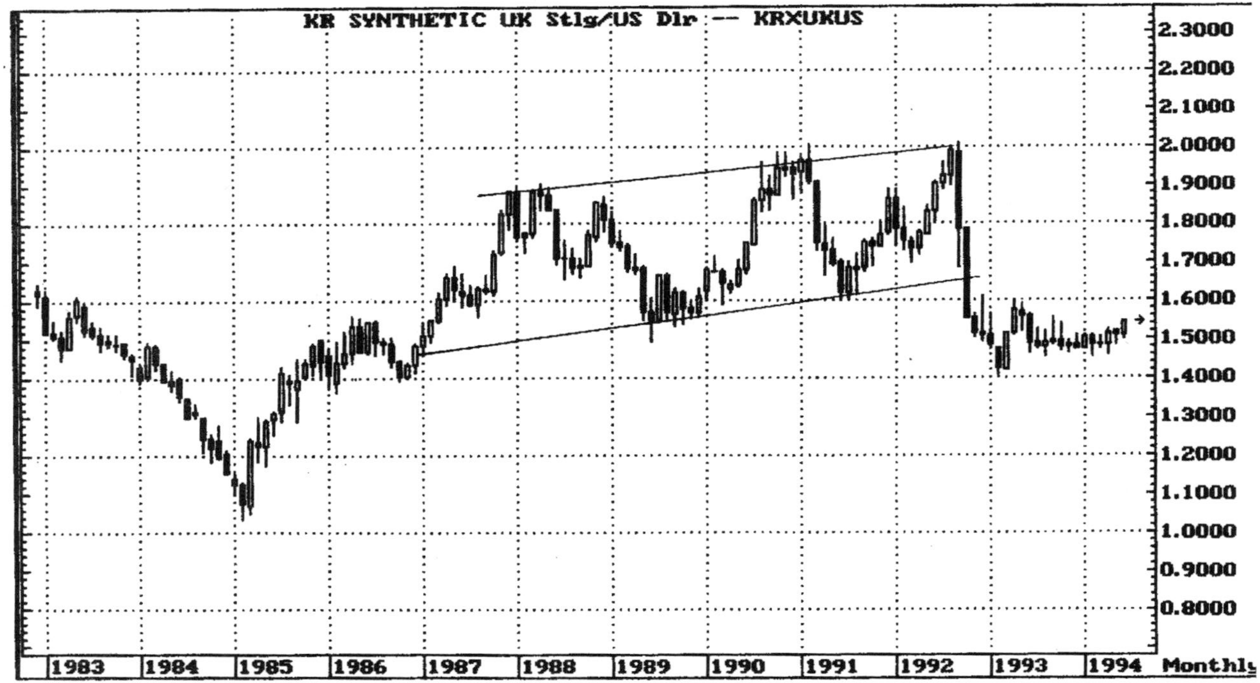

Figure 6.34. The three mountains reversal formation on the British pound/US dollar monthly chart. (*Source*: Bridge Information Systems, Inc.)

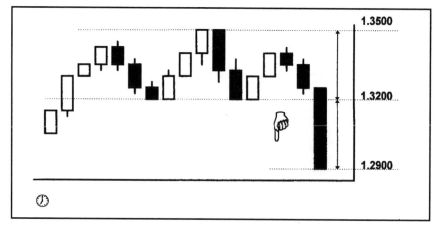

Figure 6.35. The three Buddha reversal formation.

Figure 6.36 shows a market example of the three Buddha reversal pattern.

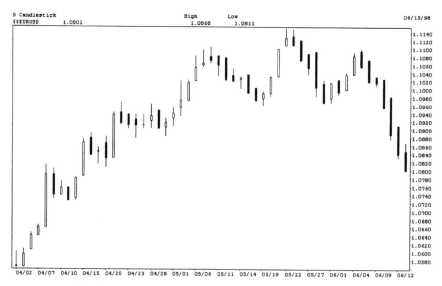

Figure 6.36. The three Buddha reversal pattern in euro/US dollar.
(*Source:* Bridge Information Systems, Inc.)

Sansen (Three Rivers or The Three River Evening Star)

The *sansen*, or the *three rivers* method, is also known as the *three river evening star*. It consists of three daily entries. The first day is a long white candlestick (a bullish move), followed by a bullish but short-ranged one-day island, and it ends with a bearish long black line (see Figure 6.37). The three rivers method replicates the exhaustion gap from the currency futures candlestick charts. The original bullish direction of the market seems to be confirmed by the bullish gap opened by the middle candlestick. The daily traders hold long overnight positions. The next day these traders are confronted by a new gap. This time, though, the gap is bearish. Traders caught with currency long positions eventually have to bail out, further fueling the bearish reversal.

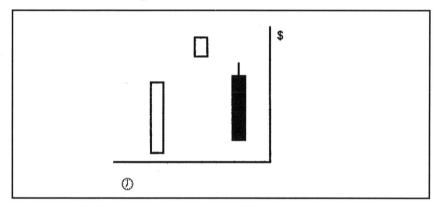

Figure 6.37. The three river evening star reversal formation.

You can see a market example of the three river evening star reversal formation in IMM Japanese yen futures in Figure 6.38.

You can also see an example of a *three river morning star* at point B.

A spinoff of the three river evening star is called, pragmatically enough, the *upside gap's two crows*. It differs in that the second and third candlesticks are both bearish, yet unable to close the exhaustion gap in the very short term. Two other versions of the three river evening star, the *evening Southern cross* and the *two crows* (see Figure 6.39), provide the *same reversal signal.*

The opposite of the three river evening star is the *three river morning star* (see Figure 6.40). The same rules apply. The three river morning star reflects a V-bottom reversal formation. Upon the completion of the third candlestick, the pattern confirms the exhaustion gap at the bottom of the trend and the signal is *to buy aggressively.*

The three river morning star is illustrated in Figure 6.41 on the IMM Japanese yen futures chart.

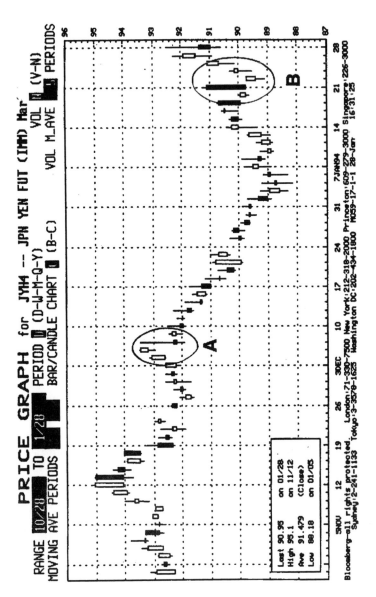

Figure 6.38. The three river evening star (A) in IMM Japanese yen futures. You can also see an example of a three river morning star at point B. (*Source:* Bloomberg.)

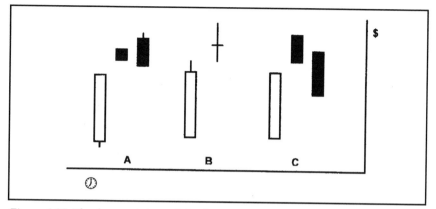

Figure 6.39. The three variations of the three river evening star: (A) the upside gap's two crows, (B) the evening Southern cross, and (C) the two crows reversal.

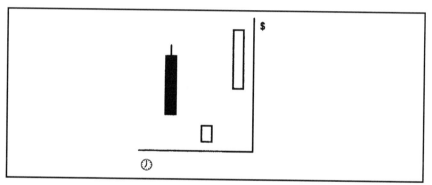

Figure 6.40. A typical three river morning star reversal formation.

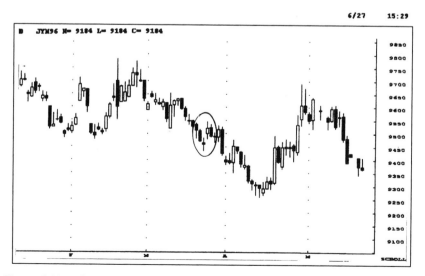

Figure 6.41. A three river morning star in IMM Japanese yen futures. (*Source:* FutureSource.)

Sanpei (Three Parallel Candlesticks)

Sanpei, or *three parallel candlesticks,* is a more complex pattern that may signal either trend continuation or trend reversal, depending on the velocity of three consecutive ranges moving in a similar direction. (By the way, the name may be redundant, because all candlesticks are parallel, but that is not a real issue.)

When the *sanpei* consists of three consecutive white candlesticks of approximately the same size that advance in approximately similar increments and the second and third candlesticks open at or above the midrange of the previous day, the reading of the pattern is that the currency is strong and will continue to rally. This bullish formation is called the *three soldiers* (Figure 6.42A).

If the second and third days post decreasingly higher highs following a long white *marubozu,* the chart signals that it is reaching the end of the up trend. So you should sell (Figure 6.42B). This pattern is known as the *red three candlestick advance block,* or *sakizumari.* (Red, of course, is the original color of the bullish candlesticks.)

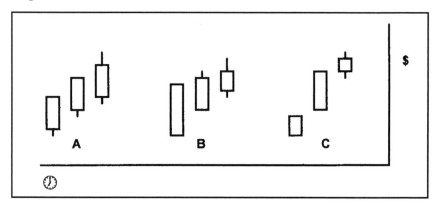

Figure 6.42. Typical white *sanpei:* (A) three soldiers, (B) red three candlestick advance block, and (C) red three candlestick star in deliberation.

Finally, if the *sanpei* is formed by a small white candlestick followed by a long white *marubozu* and ended by a white tiny-bodied candlestick, you are facing a market unsure of the next direction (Figure 6.42C). Rather than unnecessarily risk your profit, you should realize your profit. You must wait for more information because there may be a period of consolidation before the market reverses. The Japanese call this pattern the *red three candlestick star in deliberation,* or *akasansen shianboshi.*

The three soldiers formation is illustrated in Figure 6.43 on the weekly euro/US dollar chart.

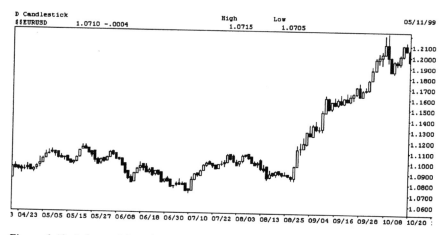

Figure 6.43. A three soldiers formation on the euro/US dollar chart. (*Source:* Bridge Information Systems, Inc.)

A bearish *sanpei* consists of three consecutive black candlesticks. Unlike the white candlestick *sanpei,* which signals both continuation and reversal, the bearish *sanpei* signals only a selloff, either as a continuation or on a breakout.

You are facing a very bearish market and you should continue to sell if:

- They have similar sizes.
- The second and third candlesticks open at or below the midrange of the previous day.
- They decline by about the same number of points.

This pattern, as seen in Figure 6.44A, is known as the *three crows.*

A variation of the three crows is *down gap three wings.* As seen in Figure 6.44B, in this bearish pattern the second-day candlestick gap is lower and the third-day candlestick open above the midrange of the second day.

Finally, a spinoff of the down gap three wings is called *simultaneous three wings*. In this case the closing of each black candlestick coincides with the opening of the following one, although the daily high and lows are different (Figure 6.44C). This pattern suggests a steep and solid sell-off.

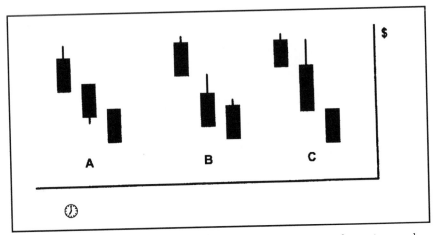

Figure 6.44. The typical black *sanpei:* (A) three crows, (B) down gap three wings, and (C) simultaneous three wings formations.

A market example of the *three crows* is provided in Figure 6.45.

Sanpo (Three Methods)

Sanpo, or *three methods,* advises a trader to occasionally pause when following a trend. In direct translation, no market goes straight up or down. Retracements must occur before the market can make new highs and new lows (Figure 6.46).

Sanpo develop as short-term consolidations with a direction opposite that of the prevailing trend. The three methods are continuation patterns, closely resembling flag formations. However, *sanpo* lack the price objective. Just like flags, these formations may be either bullish, called *rising three methods,* or bearish, called *falling three methods.*

A market example of the three falling method is provided in Figure 6.47.

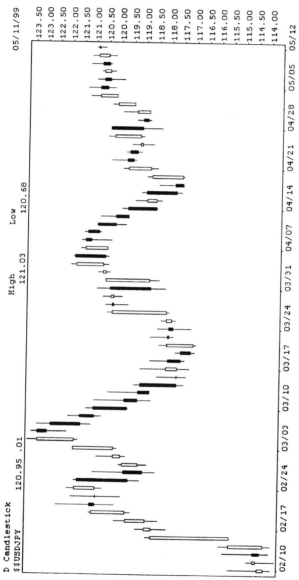

Figure 6.45. The three crows formation on the US dollar/Japanese yen chart.
(*Source:* Bridge Information Systems, Inc.)

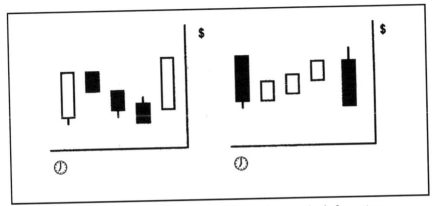

Figure 6.46. Typical rising three methods and falling three methods formations.

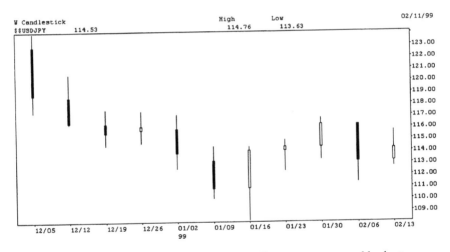

Figure 6.47. A falling method pattern on the US dollar/Japanese yen weekly chart. (*Source:* Bridge Information Systems, Inc.)

Sanku (Three Gaps)

The *sanku*, or *three gaps*, method is applicable in a market that is either steeply rising or steeply falling, where the daily limits break the trading. The theory holds that, after the third gap, the market reverses at least to the second gap. This method is rarely applicable in the currency markets because the daily limits exist only in the currency futures markets and their use is totally irrelevant. The applicability of these limits is insignificant because the cash markets, which constitute the large majority of the foreign exchange, do not have trading limits. Occasionally, this formation appears on the future charts, reflecting the steepness of the move. Figure 6.48 shows an example of a heavy selloff in the British pound futures. The selling interest on the Chicago IMM was duplicated overnight in Asia and Europe. Eventually, all gaps were filled.

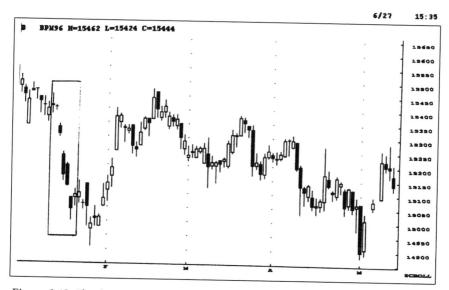

Figure 6.48. The three gaps is provided in the IMM British pound futures. (*Source:* FutureSource.)

"Wait and See" Signals

Understandably, traders like aggressive and clear trading signals: buy or sell. But the currency markets, exciting as they generally are, also generate less distinct and rather confusing moments. Some players continue to overtrade even in quiet markets, taking the additional risk of lack of liquidity. Others, versed in the less popular but equally significant "wait and see" candlestick patterns, play prudently, minimizing their brokerage costs and limiting their exposure until the market moves in a clear direction.

The candlesticks that provide the "wait and see" or neutral signals are:

- *Harami* candlestick
- *Hoshi* (star)
- *Kenuki* (tweezers)

Harami Candlestick

The *harami candlestick* is a short range that follows a long candlestick. This tiny candlestick opens and closes well within the previous day's body. The *harami* candlestick is the mirror image of the *tsutsumi* candlestick because the engulfing range occurs first. The two consecutive ranges have opposite directions, but whether they are up or down days does not matter (Figure 6.49). The market's focus remains on the activity of the first day. So the trading volume slows down until further information becomes available.

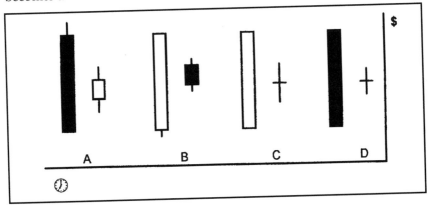

Figure 6.49. Typical *harami* candlesticks.

However, when the second day's range is a *doji* candlestick (Figures 6.49C and D), then the *harami* candlestick is called a *haramiyose candlestick*. If it occurs at an extreme level, a *haramiyose* candlestick generates a *reversal signal*. The harami candlestick is characteristic of the currency futures market because it opens at a price different from the previous day's closing.

A market example is available in the US dollar/Japanese yen daily chart in Figure 6.50.

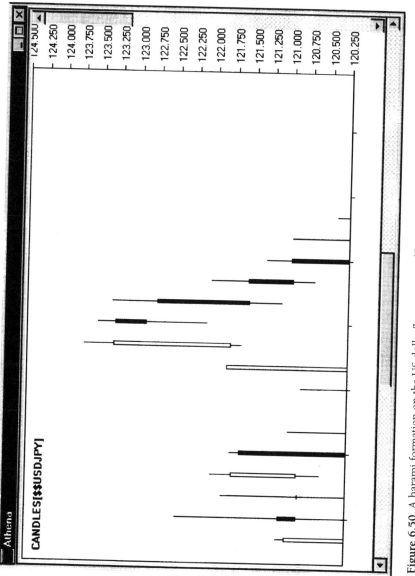

Figure 6.50. A harami formation on the US dollar/Japanese yen. (*Source:* Bridge Information Systems, Inc.)

Hoshi (Star) Candlestick

The *star candlestick*, or *hoshi*, is essentially identical in nature, if not in looks, with the *harami* candlestick. A tiny body opens and closes the following day outside the original body (see Figure 6.51) because it is unable to close the overnight gap. The presence of price gaps makes the star pattern germane to the currency futures market.

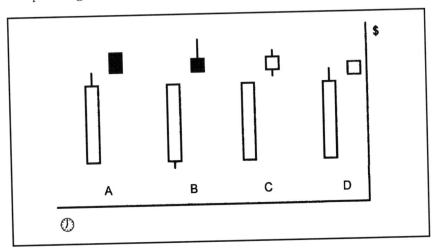

Figure 6.51. Typical star candlesticks.

At times the *hoshi* candlestick may close the overnight gap and overlap the upper shadow, but not the main body, of the first day's range, but that is not important. The direction of the two consecutive ranges is also irrelevant. The market needs more information and confidence to pursue a new direction. The reading is "wait and see." In Figure 6.51B, the star has a long upper shadow, also known as the *shooting star.*

A market example of the star candlestick is provided in Figure 6.52 on the sterling IMM futures chart.

Kenuki (Tweezers) Candlesticks

In the *kenuki,* or the *tweezers, candlesticks* formation, consecutive candlesticks have matching highs or lows. The direction of the tweezers is irrelevant, which means that they can both be bullish, bearish, or mixed.

In a rising market, a tweezers top occurs when the highs match (see Figure 6.53A), known as *kenukitenjo*. Following this development solely,

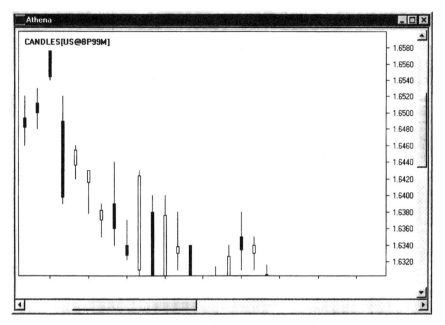

Figure 6.52. A star pattern on the sterling IMM futures chart. (*Source:* Bridge Information Systems, Inc.)

you don't have a strong direction to follow. The neutral outlook changes to reversal when two other criteria are met: (1) The formation occurs after an extended move, and (2) the second tweezers provides additional information. In Figure 6.53B, both tweezers are white and the second one is unable to make a higher high. This pattern makes a strong case for an immediate or soon-to-occur bearish reversal. Another bearish reversal signal is shown in Figure 6.53C. The second tweezers in this case is a *doji.* This type of price activity is an even more graphic example of a reversal. The Japanese call this formation *kenukiyosesen.* The only potential down side is if the tweezers do not appear at the top of the trend. As you recall from the analysis of the *doji* candlestick, a midrange *doji* is called a rickshaw man and there is no reversal.

Finally, in Figure 6.53D, the second tweezers is a black opening *marubozu* that closes below the midrange of the previous day's white candlestick. The second-day tweezers looks a lot like the bearish *kabuse* selling signal, although it does not open above the first day's high. Consequently, this pattern signals a bearish reversal.

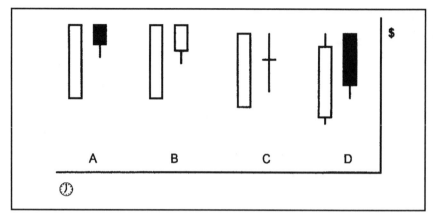

Figure 6.53. Typical tweezers tops (*kenukitenjo*) candlesticks.

A market example of a daily chart in the US dollar/Swiss franc is shown in Figure 6.54.

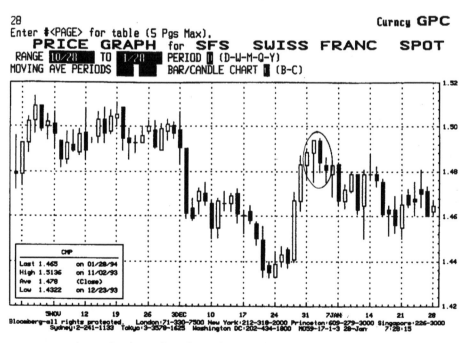

Figure 6.54. The top *kenuki* candlesticks on the US dollar/Swiss franc daily chart. (*Source:* Bloomberg Financial Services.)

In a *tweezers bottom,* two consecutive candlesticks of the same or opposite directions have similar lows (see Figure 6.55). This pattern generally provides a neutral signal, as shown in Figure 6.55A. Depending on the direction and the price structure of the second tweezers, this "wait and see" formation may forecast a bullish reversal when it appears at extreme lows.

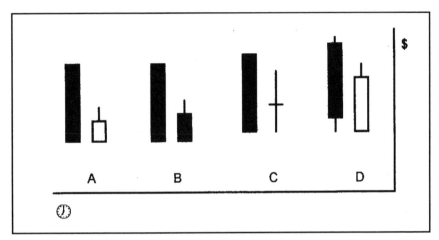

Figure 6.55. Typical tweezers bottoms candlesticks.

Figure 6.55B shows two black candlesticks, the second of which opened within the range of the first day's body, a rather undecided move. Despite the attempt to break out further on the down side, the second candlestick fails at the same low. This failure is forecasting the bottom of the down trend, and it should trigger a bullish reversal.

Figure 6.55C shows a *doji* line as the second day of the tweezers bottoms. If it occurs at the bottom of a down trend, this pattern signals a bullish reversal.

In the last type of tweezers bottom, a black candlestick is followed by a white candlestick, which opens at the previous low and closes about halfway through the body of the first candlestick. This pattern mimics the look and the meaning of the bullish reversal *kirikomi* candlestick. The only difference is that the *kirikomi* opens below the low of the first day. Market examples of the tweezers can be observed in Figure 6.56 on the euro/Canadian dollar chart.

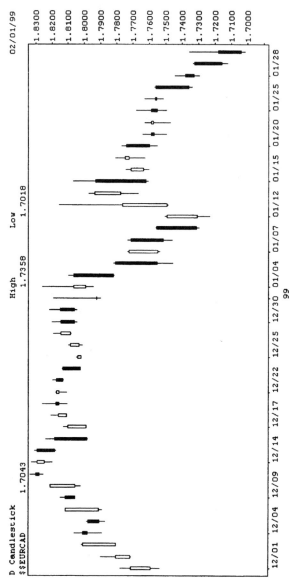

Figure 6.56. A bottom tweezers (kenuki) pattern on the euro/Canadian dollar chart. (*Source:* Bridge Information Systems, Inc.)

PART 3

QUANTITATIVE METHODS OF ANALYSIS

MOVING AVERAGES

Having analyzed all the types of charts and all the pertinent chart formations, we can focus on the quantitative methods of analysis: moving averages, oscillators, and specialty indicators. This chapter deals with moving averages.

Price Structure

A *moving average* is a lagging indicator that smooths a currency's swings. To obtain a simple average, you would calculate the arithmetic mean. To turn a simple average into a moving average, every time you add the most recent price, you deduct the oldest price. The higher the number of days you use, the more smoothing the average accomplishes.

What prices should you average? Since the electronic services track the opening, high, low, and closing prices, you can average any of these, or you can use the midpoint price (itself an average between the high and the low prices). However, most traders use the closing price because it is the most important price of the day.

Using a moving average makes it easier to look at currency movements, which sometimes resemble a seismologic needle near the San Andreas Fault. You can use a moving average by itself or as the base for other quantitative tools. Averages are popular among large numbers of traders.

Traders use three types of moving averages:

1. The simple moving average, or arithmetic mean.

2. The linearly weighted moving average.
3. The exponentially smoothed moving average.

Simple Moving Average

A *simple moving average* is calculated as the sum of a predetermined number of prices, divided by the number of entries. For example, for a 4-day moving average, add the most recent four prices and divide by four. Every time a new price is added, the fifth price is deducted, thus maintaining the same predetermined number of days: four.

$$\text{Simple moving average} = \frac{P_5 + P_4 + P_3 + P_2 + P_1 - P_5}{4}$$

Figure 7.1 shows examples of simple 5- and 15-day moving averages.

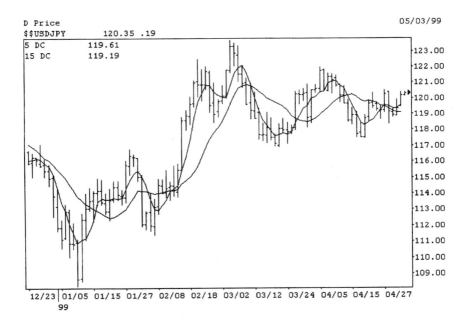

Figure 7.1. Simple 5- and 15-day moving averages on the US dollar/Japanese yen chart. (*Source:* Bridge Information Systems, Inc.)

Linearly Weighted Moving Average

A *linearly weighted moving average* assigns more weight to recent closings. This approach is arguably better for longer-term moving averages because it allocates more weight to current prices at the expense of older prices. While weighting is not significant for a seven-day moving average, it is noticeable in a 100-day moving average.

This type of moving average (see Figure 7.2) is more sensitive than the simple one and tends to turn faster. The down side is that, because of its increased sensitivity, it generates many wrong swings.

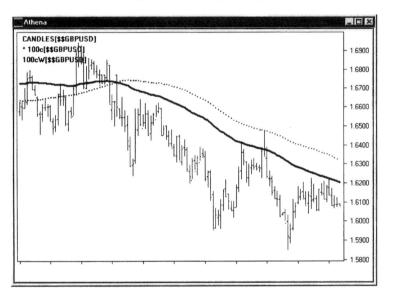

Figure 7.2. Simple 100-day moving average (dotted line) and 100-day weighted moving average on the British pound/US dollar chart. (*Source:* Bridge Information Systems, Inc.)

To calculate a linearly weighted moving average, multiply the last day's price by 1 and each following price by an increasing consecutive number up to the current closing price. Multiply by the number of days selected for the average. Then add the factors and divide the sum by the total of all weights used.

In the four-day linearly weighted moving average example, multiply the fourth day's closing price by 1, the third's by 2, the second's by 3, and the current day's by 4. You also multiply the fifth day's closing price by 0 and naturally drop it. Then you sum the new factors. The new sum thus obtained is divided by 10, which is just the sum of its multipliers 1, 2, 3, and 4.

$$\begin{array}{l}\text{Linearly weighted} \\ \text{moving average}\end{array} = \frac{(P_4 \times 1) + (P_3 \times 2) + (P_2 \times 3) + (P_1 \times 4) - (P_5 \times 0)}{1 + 2 + 3 + 4 - 0}$$

Exponentially Smoothed Moving Average

The advantage of an *exponentially smoothed moving average* is that it takes into account the previous price information of the underlying currency, in addition to assigning different weights to the previous prices. Naturally, the older the price becomes, the less weight it is assigned. To create this type of moving average, calculate the difference between the previous period's exponentially smoothed moving average and the product of the previous period's exponentially smoothed moving average and a smoothing factor.

$$EMA_C = EMA_{C-1} + ESF \times (P_C - EMA_{C-1})$$

where

EMA_C	=	current exponentially smoothed moving average
EMA_{C-1}	=	previous exponentially smoothed moving average
ESF	=	exponential smoothing factor
P_C	=	current price

$$ESF = \frac{2}{n + 1}$$

where

n = number of days in the moving average

To calculate the initial exponentially smoothed moving average, use the following formula:

$$EMA_C = P_C \times ESF + MA_C - 1 \times (1 - ESF)$$

where

EMA_C	=	current exponentially smoothed moving average
P_C	=	current price
ESF	=	exponential smoothing factor
MA_{C-1}	=	previous simple moving average

and

$$ESF = \frac{2}{n}$$

where

n = number of days in the moving average

Figure 7.3 shows an example of exponentially smoothed moving averages.

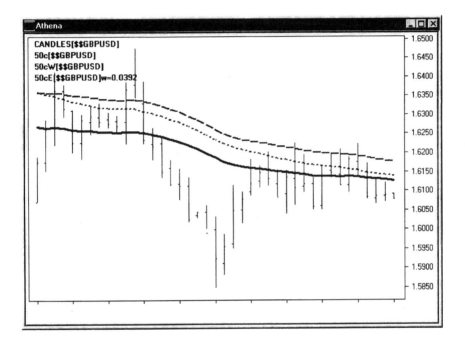

Figure 7.3. Simple 50-day exponential moving average (top), 50-day simple moving averages (middle), and 50-day weighted moving average (bottom) on the British pound/US dollar chart. (*Source:* Bridge Information Systems, Inc.)

How Many Moving Averages to Use and for What Periods

Generally, you should use one to three moving averages. However, there are no universal answers, in terms of either the number of averages to use or their duration. You must experiment and choose the number of averages that best fit the currency of your choice. Some currencies are more volatile than others, some are trending while others are not. So you must also customize the number of days in your averages to maximize the signals and minimize the frequency of false breaks.

Your market view also affects how many averages to use. For example, if all you need is a filter in a market trend, then use a single average. When an average follows below a rising price, the eventual intersection provides you with at least a warning, if not a signal, to take your profit or even reverse your strategy. This method works best when the trend is clear and your view is medium- or long-term. On the other hand, day-to-day traders want to focus on two or three moving averages and on how they interact. Three moving averages should yield an adequate number of signals. Beyond three moving averages, the chart becomes too crowded.

In terms of duration, some of the more popular periods are 4-9-18 days for the very short term, and 5-20-60 and 7-21-90 days for the medium or long term. Pragmatically, the 5-20-60 days moving average combination reflects the durations of one business week, one business month, and roughly three business months (see Figure 7.4). The 7-21-90 days set of moving averages is very popular among Japanese traders. Naturally, in addition to tracking your moving averages on a daily basis, you can use any other period, short or long term: 15 minutes, hourly, weekly, etc.

Again, don't be afraid to experiment.

Two Moving Average Combination or Double Crossover Method

On a combination of two moving averages, a *buying signal* occurs when the shorter of two averages intersects the longer one upward. See Figure 7.5. This definition allows the longer-term moving average to move either in the same direction or down with the shorter one. This structure yields many false breakouts.

The *selling signal* appears when a shorter moving average intersects a longer one downward. See Figure 7.6. Since the longer-term moving average can move either way, it can generate false selling signals.

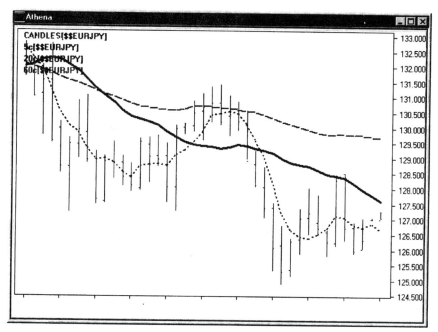

Figure 7.4. Five-day (dotted line), 20-day (continuous line), and 60-day (interrupted line) simple moving averages on the euro/Japanese yen chart. (*Source:* Bridge Information Systems, Inc.)

The two moving averages approach is also known as the *double crossover method.*

Japanese Crosses

A variation of the double crossover method that is typically used in Japan is the *cross.* Japanese technicians identify two types of intersections, or crosses: a *dead cross* and a *golden cross.*

The *dead cross* is the intersection of two consecutive moving averages as they move in *opposite directions.* Traders should disregard this type of cross, but they can use it to filter out more ambiguous intersections. Figure 7.7 demonstrates a dead cross between the 7- and 21-day moving averages in the US dollar/Japanese yen, which occurs above a 90-day moving average.

However, the intersection of two consecutive moving averages moving in the *same direction* generates a *golden cross,* which is a reliable signal that the currency will move in the same direction. See Figure 7.8. Therefore, if a rising 21-day moving average crosses above a rising 90-day moving average, this event is a bullish *golden cross,* and you may expect

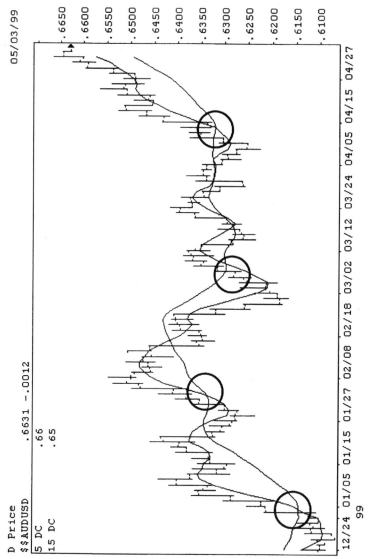

Figure 7.5. Basic buying signals generated by the intersection between the 5-day and 15-day simple moving averages on the Australian dollar/US dollar chart. (*Source:* Bridge Information Systems, Inc.)

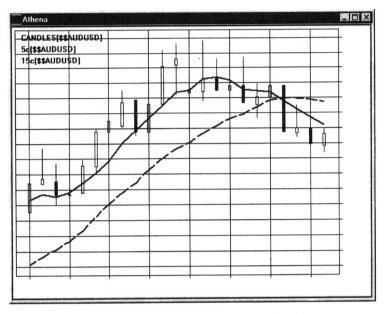

Figure 7.6. A selling signal triggered by the 5- and 15-day three moving averages displayed on the Australian dollar/US dollar chart. (*Source:* Bridge Information Systems, Inc.)

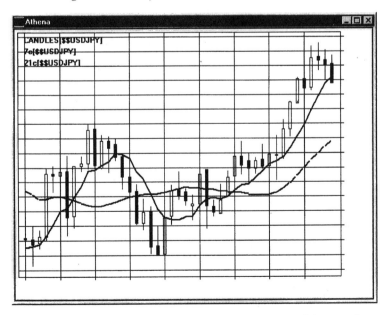

Figure 7.7. A dead cross signal triggered by the intersection of the 7- and 21-day moving averages on the US dollar/Japanese yen chart. (*Source:* Bridge Information Systems, Inc.)

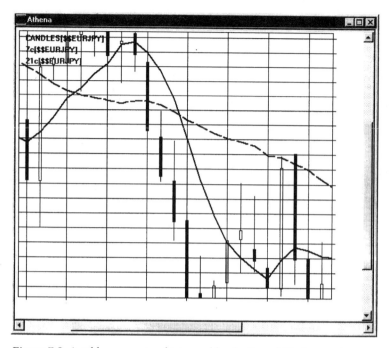

Figure 7.8. A golden cross signal triggered by the 7- and 21-day moving averages on the euro/Japanese yen chart. (*Source:* Bridge Information Systems, Inc.)

the underlying currency to rally. Better yet, you should be able to antici-pate the crossing, since the averages take some time to reach the crossing point. Generally, 21- and 90-day averages are not sensitive enough to pro-duce many false swings.

Starting with these crosses, Japanese traders identify another strong warning of reversal based on the divergence between moving averages. If the distance between two consecutive moving averages exceeds 25%, then the moving averages should start converging. Since moving averages are not leading but lagging indices, it follows that the currency must reverse to generate this change. The up side of this signal is that you see it before the currency changes direction, an unusual characteristic among moving averages. The down side is twofold. First, the 25% level repre-sents the lowest divergence to watch, but it may go even above 40% before the reversal. Second, the trigger is not seen on the averages until after the fact. That means that a sharp currency reversal is totally missed unless the move is a major trend reversal.

Three Moving Average Combination or Triple Crossover Method

A signal involving three moving averages, also called the *triple crossover method,* requires just a little more sophistication. The intersection of two shorter ones provides a warning signal, and the intersection of longer ones following the previous intersection should give a trading signal.

Let's take the example of the signals generated by the combination of the fast 4-, 9-, and 18-day moving averages. As presented in Figure 7.9, the *buying warning* occurs when the 4-day moving average, moving upward, crosses above both the 9- and 18-day averages. The *buying signal* occurs when the 9-day moving average also crosses the 18-day average upward.

The reverse is true for the selling signal. See Figure 7.10. The *selling warning* appears when the shortest average intersects downwards both the middle and the longest moving averages. This warning becomes a *selling signal* when the middle moving average falls below the longest average as well.

Donchian's 5- and 20-Day Moving Averages Method

Richard Donchian devised a very interesting method of using the moving averages, which involves general and supplemental rules. The following section quotes part of his December 1974 *Futures* article. The general rules were applied to currencies (see Figure 7.11).

A currency penetration of the moving average is confirmed only when it reaches or exceeds a full unit. The size of the unit is applied as follows:

Price Range	Unit Size	Currency
0.00 – 4.00	.01	EUR/USD, USD/CHF, GBP/USD, AUD/USD, NZD/USD, USD/CAD, etc.
40.00 – 100.00	.1	USD/JPY in 1994–1995
100.00 – 400.00	.2	USD/JPY any other times
Over 400.00	.4	USD/IDR

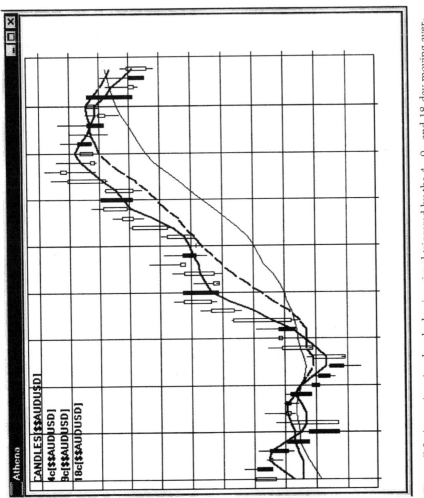

Figure 7.9. A warning signal and a buying signal triggered by the 4-, 9-, and 18-day moving averages on the Australian dollar/US dollar chart. (*Source:* Bridge Information Systems, Inc.)

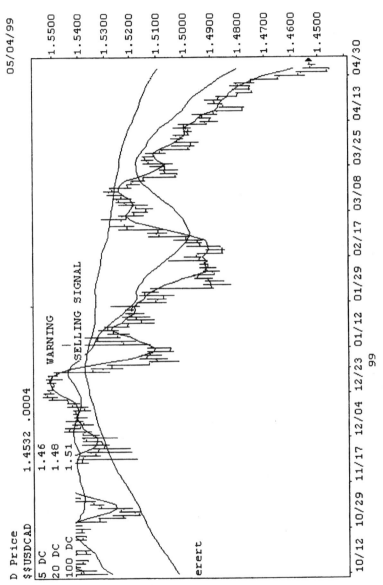

Figure 7.10. A warning signal and a selling signal generated by a combination of three moving averages in the US dollar/Canadian dollar chart. (*Source:* Bridge Information Systems, Inc.)

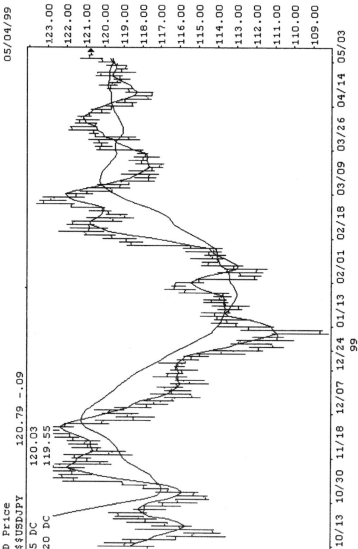

D Price
$$USDJPY   120.79 -.09
5 DC   120.03
20 DC   119.55

05/04/99

Figure 7.11. Donchian's 5- and 20-day three moving averages in the US dollar/Japanese yen chart. (*Source:* Bridge Information Systems, Inc.)

Basic Rule A

Act on all closes that cross the 20-day moving average by an amount exceeding by one full unit the maximum penetration in the same direction of any previous closing when the closing was on the same side of the moving average.

Basic Rule B

Act on all closes that cross the 20-day moving average and close one full unit beyond the previous 25 closes.

Basic Rule C

Within the first 20 days after the first day of a crossing that leads to a trading signal, reverse on any close that crosses the 20-day moving average and closes one full unit beyond the previous 15 closes.

Basic Rule D

Sensitive five-day moving average rules for closing out positions and for reinstating positions in the direction of the 20-day moving average are:

1. *Close out positions* when the currency closes below the 5-day moving average for long positions and above the 5-day moving average for short positions, by at least one full unit more than the greater of either the previous penetration on the same side of the 5-day moving average, or the maximum point of any penetration within the preceding 25 trading days. Should the range between the closing price in the opposite direction to the Rule D closeout signal be greater within the prior 15 days than the range from the 20-day moving average in either direction within 60 previous sessions, do not act on Rule D closeout signals unless the penetration of the 5-day moving average exceeds by one unit the maximum range both above and below the 5-day moving average during the preceding 25 sessions.

2. *Reinstate positions* in the direction of the basic trend (a) when the conditions in paragraph 1 are achieved, (b) if a new Rule A basic trend is given, or (c) if new Rule B and Rule C signals in the direc-

tion of the basic trend are given by closing in new low or new high ground.

3. Penetrations of two units or less do not count as points to be exceeded by Rule D unless at least two consecutive closes were on the side of the penetration when the point to be exceeded was set up.

Finally, Donchian provides *supplementary general rules*. They apply only to currency futures:

1. Action on all signals is deferred for one day except on Thursday and Friday.
2. For signals given at the close on Friday, action is taken on the open on Monday.
3. For signals given at the close on Thursday, action is taken at the Friday close.
4. When there are holidays in the middle of the week or long weekends, for sell signals use weekend rules and for buy signals defer the action by one day.

Price and Time Filters

Single moving averages are frequently used as *price and time filters,* mirroring the filters first described under the trend line breakout confirmation rules.

- As a *price filter,* a short-term moving average has to be cleared by the currency closing price, by the entire daily range, or by a certain percentage (chosen at the discretion of the trader).
- As a *time filter,* a short number of days may be used to avoid false signals.

The most popular versions of the price filters are the *envelope model,* the *high-low band, Bollinger bands,* and *Keltner channels.*

Envelope Model

In the *envelope model,* the two winding parallel lines above and below a moving average create a band bordering most price fluctuations (see Figure 7.12). To calculate the envelope model, add and subtract 2% to

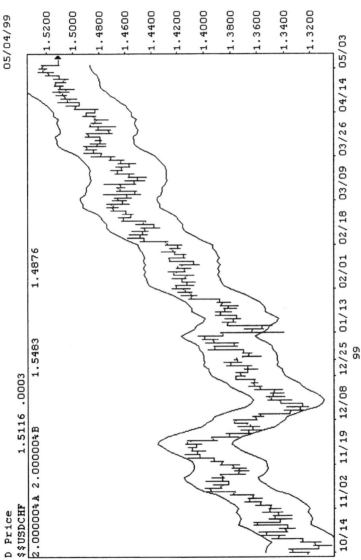

Figure 7.12. The envelope model in the US dollar/Swiss franc chart. (*Source:* Bridge Information Systems, Inc.)

and from a short-term moving average based on closing prices. Calculate the envelope bands as follows:

Upper band = MA × 1.02
Lower band = MA × .98

A *selling signal* occurs when the upper band is penetrated. A *buying signal* occurs when the lower band is penetrated. Since the signals generated by the envelope model are very short-term and occur many times against the ongoing direction of the market, speed of execution is paramount.

Although the average is not important for this method in itself, it may provide a support and resistance level. I suggest using 5 days—a business week—for your moving average.

High-Low Band

The *high-low band* consists of two moving averages of similar duration based on the high and low prices of the period. The resulting two moving averages are not parallel. The method was designed to reduce the number of false crossings.

A close above the upper band suggests a *buying signal,* and the lower band acts as your stop out level.

Conversely, a close below the lower band gives a *selling signal,* and the upper band should be used to stop yourself out.

Bollinger Bands

The *Bollinger bands,* named after John Bollinger, combine a moving average envelope with the volatility of the currency. The bands were designed to gauge whether the prices are relatively high or low. Bollinger chose standard deviation to measure the volatility because of its sensitivity to extreme deviations. The bands are plotted two standard deviations above and below a simple 20- or 21-day moving average. The Bollinger bands are calculated as follows:

The bands look a lot like an expanding and contracting envelope model (see Figure 7.13).

$$d = \frac{(P_1 - MA)^2 + (P_2 - MA)^2 + \ldots + (P_n - MA)^2}{n}$$

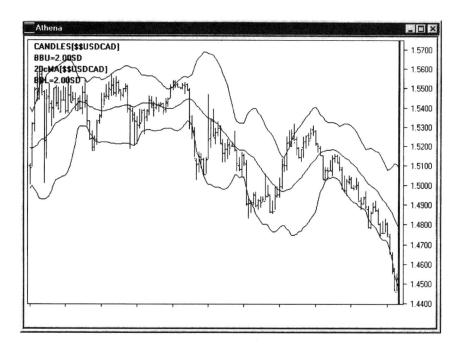

Figure 7.13. Bollinger bands in the US dollar/Canadian dollar chart. (*Source:* Bridge Information Systems, Inc.)

where

d	=	deviations from the average
P	=	price
MA	=	moving average
n	=	number of entries
σ	=	$\sqrt{d}$

where

σ	=	standard deviation

Now you can calculate the bands.

Center line = MA
Upper Band = MA + 2σ
Lower Band = MA − 2σ

When the bands converge drastically, the signal is that volatility is low and they will diverge sharply in the near future. If you are a spot or currency futures trader, you desperately need more volatility, so this type of signal is helpful. The down side is that this method doesn't provide the direction when the volatility is low.

Conversely, when the bands are far apart it means that the volatility is too high and it needs to slow down. In this scenario, you already know what the original direction was, so that the direction of reversal is obvious. You must only worry about the timing of the reversal (see Figure 7.14).

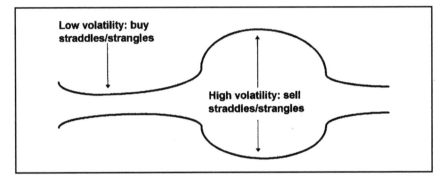

Figure 7.14. An example of Bollinger bands generated signals for option traders.

The Bollinger bands are an excellent tool for currency options traders. Unlike spot or currency futures traders, currency options traders welcome low volatility. Low volatility enables them to sell or write naked options or strategies, as well as providing them with the opportunity to buy medium-term (two- to three-month) strangles or straddles. *Strangles* are option strategies in which you buy a similar number of calls and puts with different strike prices. *Straddles* are similar to strangles except that they have the same strike price. Strangles and straddles are low-risk strategies for buying volatility without needing to know the direction of the market. All they need is a move in the underlying currency significant enough to exceed the premiums they paid for both the calls and the puts. After the market makes its move, they can either sell their currency options position or, since they know the direction of the market, just the losing side of your strategies, either the calls or the puts. The reason for the latter choice is that in a long straddle or strangle, they should make money in the calls and lose in the puts, or vice versa. *Legging out* of one of these options strategies means that you can change your original volatility position to a directional position.

An additional signal from the Bollinger bands is a succession of two tops, with one occurring outside the band, followed by one inside. This is a selling signal. Conversely, a succession of two bottoms, with the first one below the lower band and the following one above is a buying signal.

Keltner Channels

The *Keltner channels* attempt to identify minor trends in their incipiency by comparing current prices against previous prices. The absence of new lows signals an up trend, while the absence of new highs points to a down trend (see Figure 7.15). In addition, you can use the *Minor-Trend Rule,* which asserts that the minor trend is:

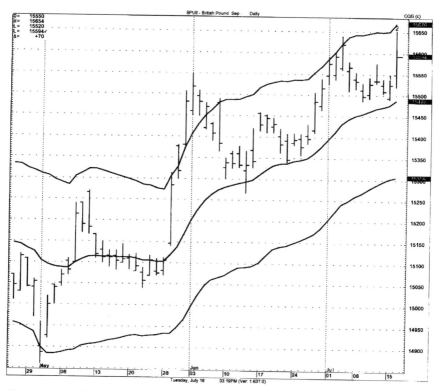

Figure 7.15. Keltner channels in the British pound/US dollar.
(*Source:* CQG. ©Copyright CQG INC.)

- Bullish when the daily trend sells above its recent high.
- Bearish when the daily trend sells below its recent low.

Like all the other channel studies, the Keltner channels consist of a moving average for the center line and two channel lines for the upper and lower bands. The channel lines are calculated by adding to and subtracting from the center line the product of a constant and the average true range of each entry.

To calculate the Keltner channels, use the following formulas:

$$CL = \frac{EMA\ (n-1) + P}{n}$$

where

CL	=	center line
EMA	=	exponential moving average
n	=	number of entries
P	=	price

$$ATR = \frac{R\ (n-1) + (H-L)}{n}$$

where

ATR	=	average true range
r	=	range
n	=	number of entries
H	=	high price
L	=	low price

Upper band = $(CL + ATR) \times C$
Lower band = $(CL - ATR) \times C$

where

CL	=	center line
ATR	=	average true range
c	=	constant

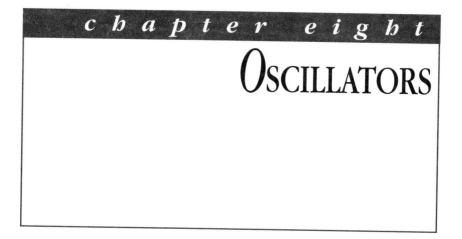

c h a p t e r   e i g h t

OSCILLATORS

As the market gyrates, prices tend to overshoot, to overextend. Oscillators are *derived* from the underlying currency to provide signals regarding these overbought and oversold conditions. They come in particularly handy during trading ranges or at the beginning of new trends.

General Rules

As we present the oscillators in more detail, three rules will become apparent:

1. *The signals are most useful at the extremes of their scales.* Keep a basic thing in mind: Oscillators may be built on open positive/negative scales around zero (the equilibrium level) or on close scales between 0 and 100. When using oscillators designed on open scales, you must identify "extreme levels" (see Figure 8.1), relative to the equilibrium level or the zero line. To this end, check the historical behavior of a specific oscillator for each of the underlying currencies in which you are interested for several long-term and short-term periods. Currencies have different degrees of volatility, and they can be free floating, semipegged, and fully pegged.

2. *Crossings of the equilibrium line (where applicable) and crossings between oscillator lines usually generate direction signals.* Temporary penetrations of the equilibrium, like temporary penetrations of the moving averages, should be disregarded. In Figure 8.1, the ample bullish crossover A signals a bullish currency, but the meek bearish crossing B should be disregarded since the currency is in a sideways pattern.

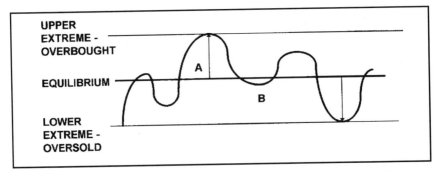

Figure 8.1. Identifying the "extreme levels" relative to the equilibrium level or the zero line.

3. *Warnings and/or signals are triggered when a divergence occurs between the price of the underlying currency and the oscillator.* When both the underlying currency and its oscillator move in the same direction and they both have approximately similar tops and bottoms, they are said to be *in gear* (see Figure 8.2). This development does not

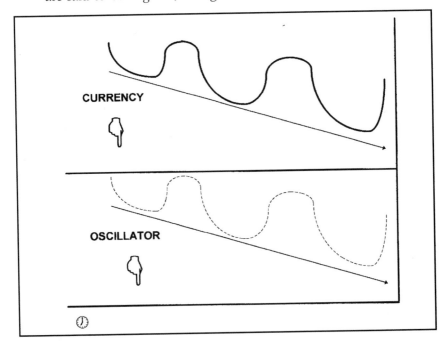

Figure 8.2. A currency and its oscillator in gear.

help you much, except to reinforce the fact that the currency will continue the current direction or trend.

If the underlying currency and its oscillator diverge, you get currency reversal signals. When the descending tops of a bearish currency are followed by ascending tops in the oscillator, you are facing a *bullish divergence* (see Figure 8.3). In this situation, the read is that the currency will stage a bullish reversal. Figure 8.4 shows a market example of bullish divergence between the euro/US dollar and the oscillator.

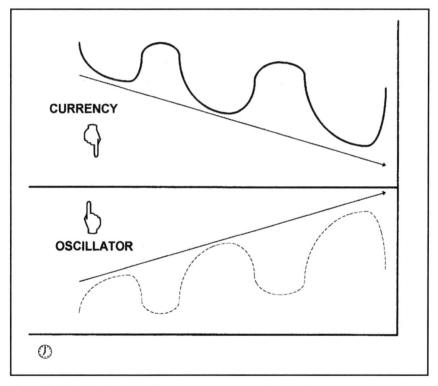

Figure 8.3. Bullish divergence between a currency and its oscillator.

If the ascending tops of a bullish currency are followed by descending tops in the oscillator, then the up trend is not supported. This *bearish divergence* gives you a selling signal (see Figure 8.5).

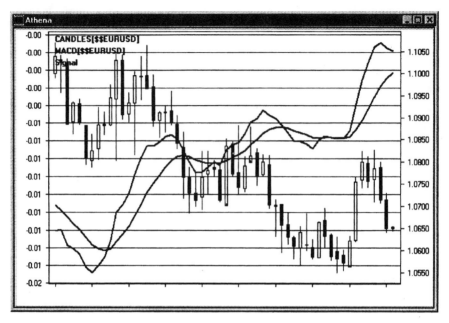

Figure 8.4. Example of a bullish divergence between the euro/US dollar and the MACD oscillator. (*Source:* Bridge Information Systems, Inc.)

Figure 8.6 shows a market example of bearish divergence between the US dollar/Japanese yen and the MACD.

When using the divergences, you must be careful in terms of timing. Identifying the divergences does not automatically trigger a reversal. Since divergences may linger, you must use a trigger from a different source.

In terms of the time periods for the underlying currency, start with a long-term period, such as weekly. Continue with daily, and end with whatever short-term period you like best: hourly, 15 minutes, etc.

The analysis of the following oscillators includes all the pertinent formulas. Some of these formulas are very easy and some are complex. You don't need to really remember them. It is, however, important to know how they work so that you can better understand both their value and their limitations.

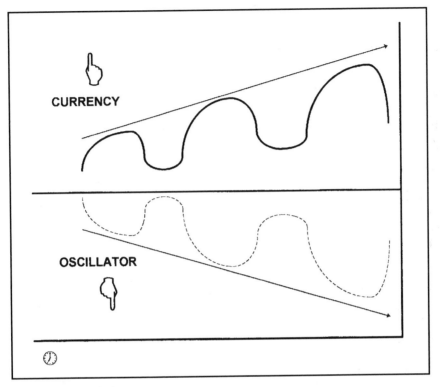

Figure 8.5. Bearish divergence.

Momentum

Momentum is a fundamental oscillator designed to measure the rate of price change, not the actual price level. This oscillator consists of the net difference between the current closing price and the oldest closing price from a predetermined period. The formula for calculating the momentum is:

$$M = CCP - OCP$$

where

M	=	momentum
CCP	=	current closing price
OCP	=	old closing price for the predetermined period

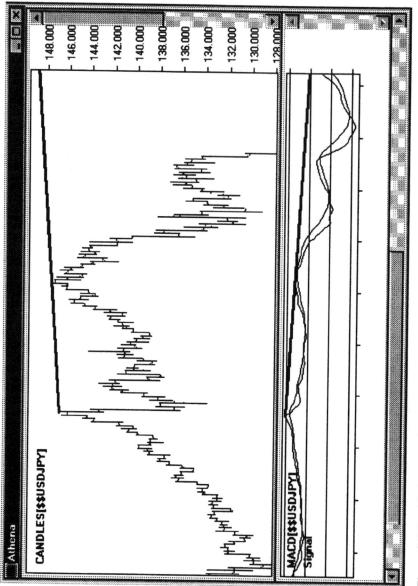

Figure 8.6. Bearish divergence between the US dollar/Japanese yen and the MACD. (*Source:* Bridge Information Systems, Inc.)

The values thus obtained are either positive or negative numbers on an open scale, and they are plotted around the zero line. At extreme positive values, momentum suggests an overbought condition, whereas at extreme negative values, an oversold condition is indicated. See Figure 8.7. (An open scale requires the trader to fine-tune the historical information in order to gauge the magnitude of these extremes.)

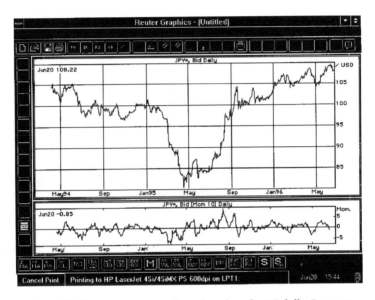

Figure 8.7. The momentum oscillator based on the US dollar/Japanese yen daily chart. (*Source:* Reuters.)

The signals triggered by the crossing of the zero line remain in effect. However, they should be followed only when they are consistent with the ongoing currency trend. In other words, buy on a positive penetration *only if* the currency is already bullish, and sell on a negative crossover *only if* the currency has already turned bearish. Momentum generates its trading signals ahead of the currency.

In terms of time frame, the shorter the number of days included in the calculations, the more responsive the momentum becomes to short-term fluctuations and false signals, and vice versa. Typically, 10 periods are included in the momentum calculation.

Rate of Change (ROC)

The *rate of change* (see Figure 8.8) is a close version of the momentum oscillator. The difference is that, while the momentum's formula is based on subtracting the oldest closing price from the most recent, the ROC's formula is based on *dividing* the oldest closing price into the most recent one.

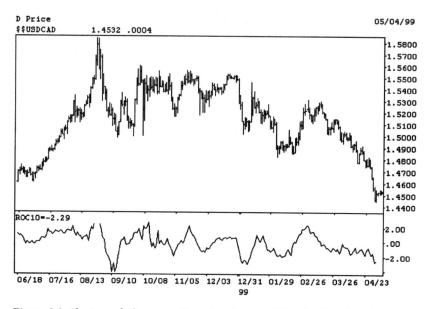

Figure 8.8. The Rate of Change oscillator based on the US dollar/Canadian dollar chart. (*Source:* Bridge Information Systems, Inc.)

The formula for calculating the ROC is:

$$ROC = \frac{CCP}{OCP} \times 100$$

where

ROC = rate of change
CCP = current closing price
OCP = oldest closing price for the predetermined period

The default number of periods is 10, although you can use any number of days.

Similar to the momentum oscillator, the ROC is also designed on an open scale, but the equilibrium line occurs at 100 rather than 0. The standard momentum rules of interpretation apply.

Commodity Channel Index (CCI)

The *Commodity Channel Index (CCI)* was developed by Donald Lambert, primarily for the futures market, to identify the start and the end of cycles. CCI is calculated as the difference between the mean price of the currency and the average of the mean price over a period of 5 to 20 days. This oscillator swings on an open scale (see Figure 8.9).

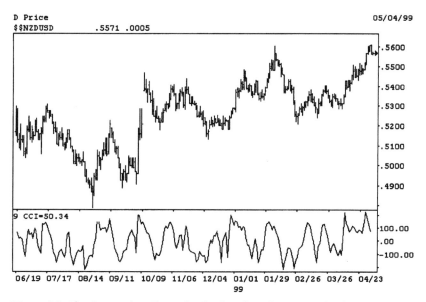

Figure 8.9. The Commodity Channel Index based on the New Zealand dollar/US dollar chart. (*Source:* Bridge Information Systems, Inc.)

The formulas for calculating the CCI are:

$$P = \frac{H + L + C}{3}$$

$$MA_n = \frac{P_1 + P_2 + P_3 + \ldots + P_n}{n}$$

$$MD = \frac{|\sum_{i=1}^{n} P_i - MA_n|}{n}$$

$$CCI = \frac{(P - MA_n)}{.015 \times MD}$$

where

P	=	average daily price
H	=	high price
L	=	low price
C	=	closing price
MA	=	moving average
MD	=	median deviation

Specific Trading Signals for the CCI

The area above +100 line is the overbought territory. Conversely, the area below −100 line is the oversold territory.

   Buying signals are generated in the following situations:

1. CCI exceeding the lower (−100) line toward the equilibrium line at zero.
2. Bullish crossings above the zero line.
3. Bullish divergence.

   Selling signals occur in the following situations:

1. CCI falling below the overbought line (+100) toward the equilibrium line at zero.

2. Bearish crossings below the zero line.

Stochastics

The stochastic oscillator was designed by George Lane, who noticed that:

- As the market approaches the end of an up trend, the closes tend to approach the daily highs.
- Toward the end of a down trend, the closes tend to draw near the daily lows.

This oscillator is versatile and reliable.

Stochastics consist of two lines called %K and %D. Visualize %K as the plotted instrument and %D as its moving average. %K line is more sensitive than %D line, and the %D line triggers the trading signals.

The formulas for calculating the stochastics on a 9-day period are as follows:

$$\%K = \frac{CCL - L_9}{H_9 - L_9} \times 100$$

where

CLC = current closing price
L_9 = the lowest low of the past 9 days
H_9 = the highest high of the past 9 days

and

$$\%D = \frac{H_3}{L_3} \times 100$$

where

H_3 = the 3-day sum of $(CCL - L_9)$
L_3 = the 3-day sum of $(H9 - L_9)$

The resulting lines are plotted on a 1- to 100-scale (see Figure 8.10). In its original form this oscillator is known as the *fast stochastics*.

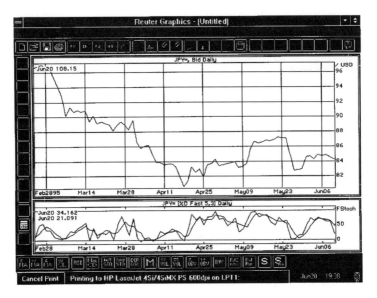

Figure 8.10. The fast stochastics study based on the US dollar/Japanese yen chart. (*Source:* Reuters.)

Specific Trading Signals for Stochastics

Extreme Values

The 70% value is used as an overbought warning signal, and the 30% value is used as an oversold warning signal. In foreign exchange, the most reliable bullish reversal signals occur under 5%, and the bearish reversal signals come into play above 95%. That is different from Lane's prescribed levels of 15% on the buying side and 85% on the selling side.

Beware of the timing of the reversal. A reading of zero obviously shows that the currency is oversold and that it will turn bullish. On a daily chart, the reversal commonly occurs the next day. However, the reversal may be delayed for a couple of days. Why? In the formula for calculating %K, the numerator is $CCL - L_9$. A current reading of zero means that today the currency closed at the lowest level in 9 days, a bearish situation. If tomorrow the %K line is at, say, 5%, the currency has not necessarily reversed, but rather closed marginally above the low of the day. To maximize your success, wait until the %K and %D lines turn downward above 95% before selling, and until the %K and %D lines turn upward below 5% before buying. The reversal will occur but, unless you have deep pockets, you should hold the reins tightly to avoid losses.

The Intersection of the %D and %K Lines

The intersection of the %D and %K lines generate further trading signals. There are two types of such intersections: left-hand and the right-hand crossings. Beware of temporary intersections. Disregard short-term crossovers where the %K line crosses above the %D line and then below it, and vice versa.

The *left-hand crossing* occurs when the %K line crosses prior to the peak of the %D line. See Figure 8.11. This type of intersection is the less desirable of the two.

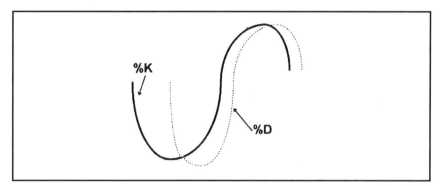

Figure 8.11. The left-hand crossing in stochastics.

The *right-hand crossing* occurs when the %K line intersects after the peak of the %D line. See Figure 8.12. Lane suggests using only the right-hand intersection trading signals. Although traders like this type of crossing to occur at the extremes of the stochastics scale, the signal remains valid at any level.

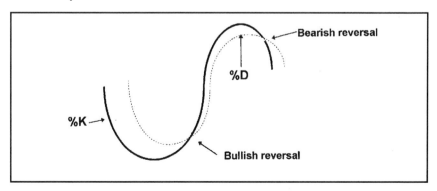

Figure 8.12. The right-hand crossing in stochastics.

Figure 8.13 shows you three market examples of the *right-hand crossing* in stochastics.

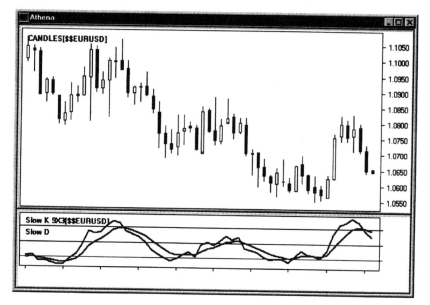

Figure 8.13. Examples of right-hand crossovers on the slow stochastics on the euro/US dollar chart. (*Source:* Bridge Information Systems, Inc.)

As in the case of all the oscillators, the divergence between the direction of the stochastics and the direction of currency price generates significant trading signals.

The Hinge

In Figures 8.11, 8.12, and 8.13, as the %K or %D line approaches reversal, its velocity slows down. This development is called the *hinge*. Figure 8.14 shows these patterns.

Slow Stochastics

Some traders feel that the %K line may be too sharp at the edges and too sensitive. To improve their analysis, traders also use slow stochastics, a version they believe provides more accurate signals. The new slow %K line replaces the original %D line. The new %D line formula is calculated from the new %K line. The result is a very smooth pair of oscillators. See Figure 8.15. Use the same analysis as for the fast stochastics.

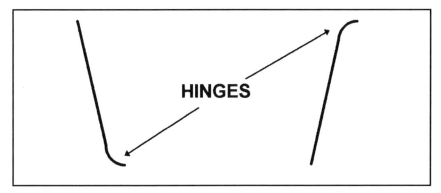

Figure 8.14. Hinges.

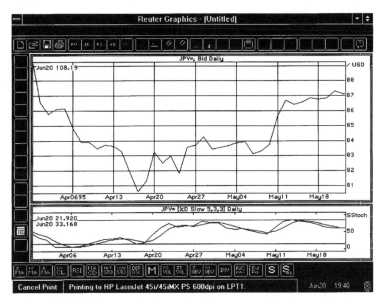

Figure 8.15. Slow stochastics based on the US dollar/Japanese yen daily chart. (*Source:* Reuters.)

Williams %R

The *Williams %R* is a version of the stochastics oscillator, which was designed by Larry Williams. The difference between the high price of a predetermined number of days and the current closing price is divided by the total range. This oscillator is plotted on an inverted 0 to 100 scale.

$$\%R = \frac{H_9 - CCL}{H_9 - L_9} \times 100$$

where

CCL	=	current closing price
H_9	=	the highest high of the past 9 days
L_9	=	the lowest low of the past 9 days

Therefore, bullish reversal signals occur under 80%, and, conversely, bearish signals appear above 20%. The interpretations are similar to those discussed under stochastics. See Figure 8.16.

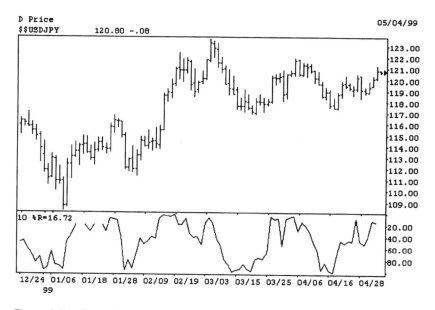

Figure 8.16. The Williams %R oscillator based on the US dollar/Japanese yen chart. (*Source:* Bridge Information Systems, Inc.)

The Relative Strength Index (RSI)

The *Relative Strength Index* (see Figure 8.17) is yet another popular oscillator, devised by Welles Wilder. The RSI measures the relative changes between the higher and lower closing prices, and it is plotted on a 0 to 100 scale. The formulas for calculating the RSI are:

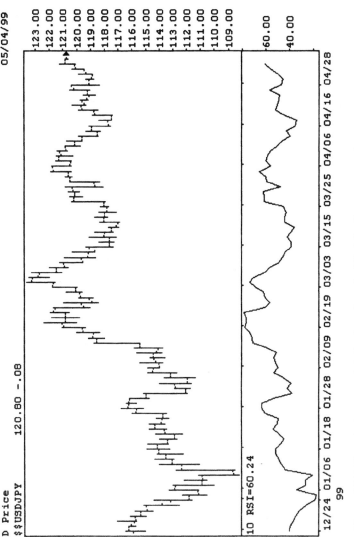

Figure 8.17. The relative strength index based on the US dollar/Japanese yen chart. (*Source:* Bridge Information Systems, Inc.)

$$RSI = 100 - \frac{100}{1 + RS}$$

and

$$RS = \frac{\text{Average of } n \text{ days' closes UP}}{\text{Average of } n \text{ days' closes DOWN}}$$

where

RSI = relative strength index
n = predetermined number of days

The original number of days used by Wilder was 14 days, or half a lunar cycle. Currently, a 10-day period, the business version that allows for weekends, is more popular.

The 70% and 30% levels are used as warning signals. As you can see in Figure 8.18, the warning levels may provide helpful trading signals. If the RSI makes a double top formation, with the first top at or above and the second below the 70%, you get a selling signal when the oscillator dips through the first trough at point A. Conversely, an RSI double bottom, with the first trough at or above 30% and the second not exceeding it, is a buying signal when the oscillator ascends above the first top at point B.

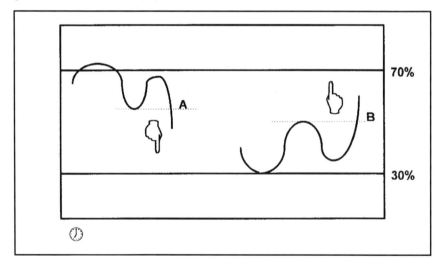

Figure 8.18. Trading signals generated by the RSI at the warning levels.

Values above 90% indicate an overbought condition, which should trigger a *selling signal*. An RSI under 10% reflects an oversold condition and a *buying signal*. Wilder identified the RSI's forte as its divergence versus the underlying price. Some traders use only the numerical values. Beware of false reversal signals when trading solely on the extreme levels. The RSI can spend fairly long periods of time in the overbought or oversold territories before the currency reverses.

Moving Averages Oscillators

As oscillators, the values of two consecutive moving averages are subtracted from each other. See Figure 8.19. This basic oscillator uses simple moving averages and fluctuates on an open scale around the zero line.

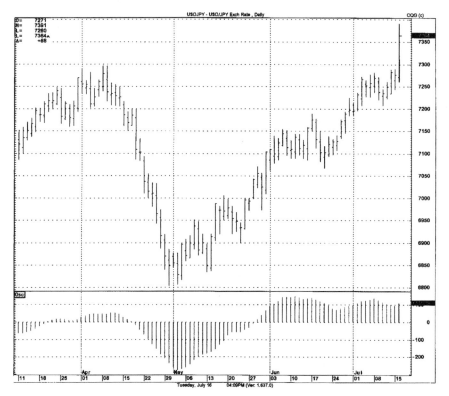

Figure 8.19. The moving average oscillator based on the US dollar/Japanese yen daily bar chart. (*Source:* CQG. ©Copyright CQG INC.)

Moving average oscillator = $MA_1 - MA_2$

Trading Signals for Moving Averages Oscillators

1. When the oscillator line intersects the zero line upwards, a *buying signal* occurs, and vice versa.
2. A divergence between the moving averages oscillator and the currency triggers a *warning signal*.
3. In addition, minor trends may be spotted within the major trends.

Moving Average Convergence/Divergence (MACD)

Gerald Appel developed an oscillator built on exponentially smoothed moving averages called the *moving average convergence/divergence (MACD)*. The MACD consists of the difference between two exponential moving averages on 12-day and 26-day, respectively, that are plotted on an open scale against the zero line. See Figure 8.20. The zero line repre-

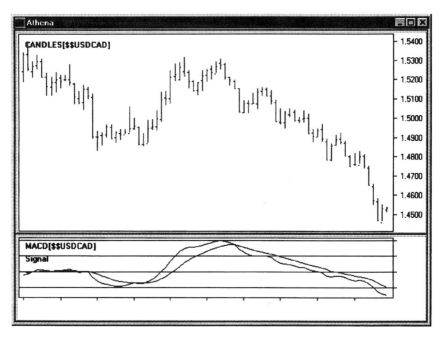

Figure 8.20. Moving average convergence/divergence (MACD) based on the US dollar/Canadian dollar chart. (*Source:* Bridge Information Systems, Inc.)

sents the times when the values of the two moving averages are identical. The MACD can be represented on the chart as a line or as a histogram. In addition to this line, an additional 9-day exponential moving average acts as a *trigger* or *signal* line. MACD is fairly popular among traders.

Moving average oscillator = $EMA_{12} - EMA_{26}$

where

EMA_{12} = 12-day exponentially smoothed moving average
EMA_{26} = 26-day exponentially smoothed moving average

Naturally, you can use moving averages of different durations.

Trading Signals for the MACD

Buying Signals

Buying signals occur when:

1. MACD rises above the zero line.
2. MACD rises above the trigger line (see Figure 8.20).
3. A bottom in the underlying currency is not "in gear" with a rising MACD (bullish divergence).

Selling Signals

Selling signals occur when:

1. MACD falls below the zero line.
2. MACD falls below the trigger line.
3. A top in the underlying currency is not "in gear" with a falling MACD (bearish divergence).

If you prefer to represent MACD as a single line, use the *MACD oscillator* (see Figure 8.21). The MACD oscillator is the difference between the MACD line and the 9-day exponentially smoothed moving average.

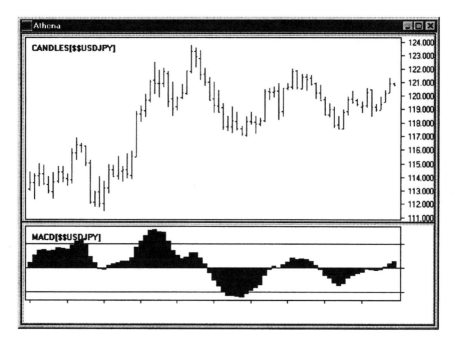

Figure 8.21. Moving average convergence/divergence histogram based on the US dollar/Japanese yen chart. (*Source:* Bridge Information Systems, Inc.)

Oscillator Combinations

Traders use different combinations of technical tools in their daily trading and analysis. Some of the more popular combinations are shown in Figures 8.22 and 8.23.

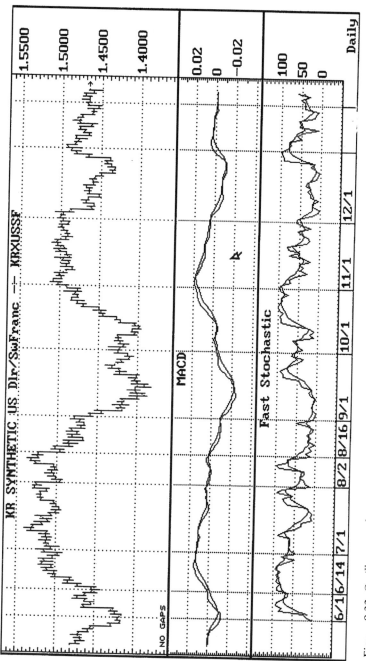

Figure 8.22. Oscillator combinations in US dollar/Swiss franc. (*Source:* Bridge Information Systems, Inc.)

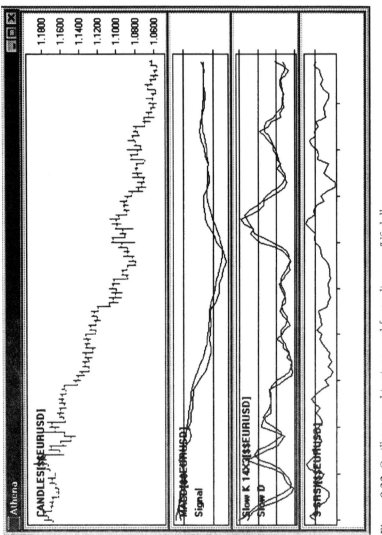

Figure 8.23. Oscillator combinations used for trading euro/US dollar. (*Source:* Bridge Information Systems, Inc.)

256

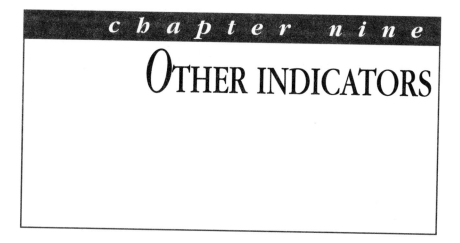

chapter nine

OTHER INDICATORS

Besides moving averages and oscillators, other quantitative tools are available for technical analysis. For the lack of better terminology, they are grouped under the general heading of *other indicators*. These indicators are powerful and increasingly popular.

Parabolic System, or SAR

Welles Wilder, who also devised the relative strength index, designed this technical tool to supplement the inadvertent gaps of the other trend-following systems. The *parabolic system* is a stop-loss system, based on price and time, that can be used in conjunction with the oscillators.

The name of the system is derived from its parabolic shape, which follows the price gyrations and is represented by a dotted line. When the parabola is placed under the price, it suggests a long position. Conversely, a parabola above the currency price indicates a short position. See Figure 9.1.

The merit of this system is that it catches new directional moves in their infancies, and some of these moves turn into more significant trends. If the new direction or trend fails, the currency intersects the parabola, thus generating the *stop and reverse*, or SAR, signal. A new parabola appears on the opposite side. The stop moves daily in the direction of the new trend. The built-in acceleration factor pushes the SAR to catch up with the currency price.

SAR performs best in trending markets, but it is less reliable in trading ranges.

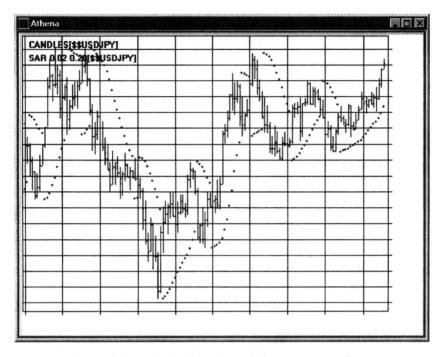

Figure 9.1. The parabolic study (SAR) on the US dollar/Japanese yen chart.
(*Source:* Bridge Information Systems, Inc.)

To calculate SAR, start at an extreme high or low price, and use an acceleration factor to multiply the extreme prices until the position is stopped out. The acceleration factor starts at 0.02 and it can extend up to 0.2.

$$SAR_C = SAR_{C-1} + AF \times (EP - SAR_{C-1})$$

where

SAR_C = current SAR
SAR_{C-1} = previous SAR
EP = extreme point
AF = acceleration factor

Directional Movement Index (DMI)

The *directional movement index* provides a signal of trend presence in the market. The line simply rates the price directional movement on a scale of 0 to 100. The higher the number, the better the trend potential of a movement is, and vice versa.

The DMI consists of three lines: the ADX line (see Figure 9.2), which measures the trend incidence, and two lines reflecting the positive and negative values of the directional indicator line (see Figure 9.3).

Calculating the DMI entails several steps:

First, calculate the *directional movement*. This is the difference between the current day's range and the previous day's range or the part of the current day's range that falls outside the previous day's range. See Figure 9.4.

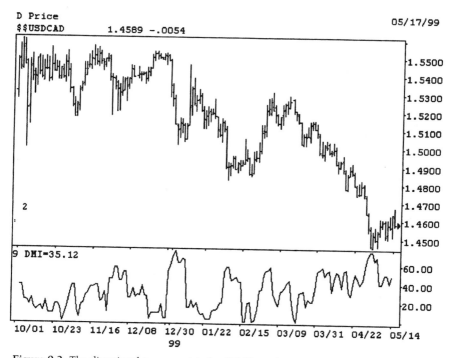

Figure 9.2. The directional movement index (DMI) on the US dollar/Canadian dollar chart. (*Source:* Bridge Information Systems, Inc.)

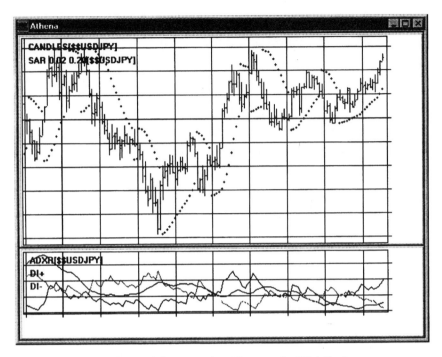

Figure 9.3. The ADX line and the +DI and –DI lines on the US dollar/Japanese yen chart. (*Source:* Bridge Information Systems, Inc.)

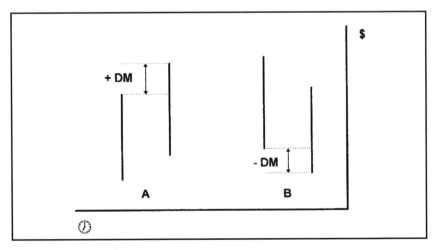

Figure 9.4. Directional movement.

When the range of a day engulfs the range of the previous day, the larger range is called an *outside day*. See Figure 9.5. (This is similar to the *tsutsumi* pattern discussed in Chapter 6.)

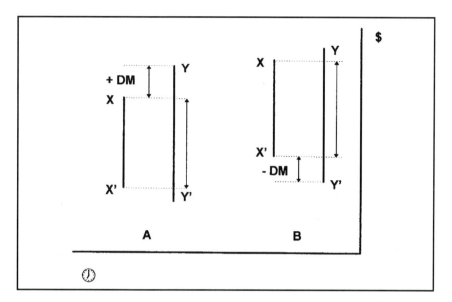

Figure 9.5. Outside days.

Conversely, a daily range that occurs within the boundaries of the previous daily range is called an *inside day*. See Figure 9.6. (This pattern is similar to the *harami* pattern in candlestick analysis.)

Second, calculate the *true range*, which is the greatest of the following:

1. $H_C - L_C$
2. $H_C - C_{C-1}$
3. $L_C - C_{C-1}$

where

H_C = current high
L_C = current low
C_{C-1} = previous closing

Figure 9.7 shows an example of the average true range.

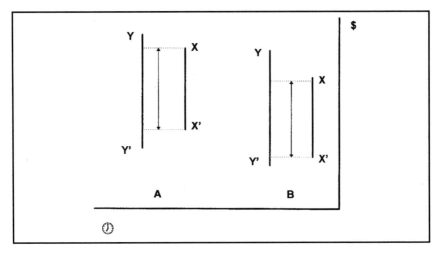

Figure 9.6. Inside days.

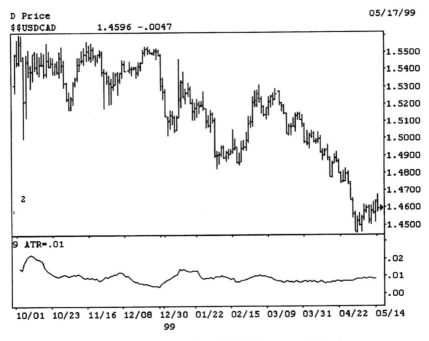

Figure 9.7. The average true range on the US dollar/Canadian dollar chart.
(*Source:* Bridge Information Systems, Inc.)

You can now calculate the *directional indicator (DI)*.

$$+ DI = \frac{+ DM}{TR} \qquad - DI = \frac{- DM}{TR}$$

$$+DI_{14^c} = \frac{DM_{14^{c-1}} + DM_{14^c} - \dfrac{DM_{14^{c-1}} + DM_{14^c}}{14} + DM_1}{TR_{14}} \times 100$$

$$-DI_{14^c} = \frac{DM_{14^{c-1}} - DM_{14^c} - \dfrac{DM_{14^{c-1}} - DM_{14^c}}{14} + (-DM_1)}{TR_{14}} \times 100$$

where

DI_{14^c} = current directional index for 14 days
DM_{14^c} = current directional movement for 14 days
$DM_{14^{c-1}}$ = yesterday's directional movement for 14 days
DM_1 = directional movement for today

Based on this data, you can now calculate the *directional movement index (DX)*.

$$DX = \frac{+ DI_{14} - (- DI_{14})}{+ DI_{14} + (- DI_{14})}$$

During periods of high volatility, you can use the *average directional movement index (ADX)*, which is calculated as follows:

$$ADX_C = \frac{ADX_{C-1} \times (n - 1) + DX_C}{n}$$

where

ADX_C = current directional index
ADX_{C-1} = previous directional index
DX_C = yesterday's directional movement for 14 days
DM_1 = directional movement for today

Again, Wilder originally used a period of 14 days, which is half the lunar cycle. You may want to experiment with the 10-day period, which is similar to the original period but allows for nontrading weekends. The number of days for these calculations is ultimately up to you. The preceding formulas are not everybody's cup of tea. However, knowing how the indicator works is important, not memorizing the formulas. The electronic services have the entire package prepared for you.

The higher the ADX, the more trend the currency has. Conversely, a reading below 25 is a strong warning to avoid trading.

The intersections between +DI and −DI should provide trading signals. A high number of crossings denote a low ADX, while failed attempts to cross over reflect a high ADX. Also, the change of direction of the ADX when it floats above both the DI lines signals a trend reversal. The currency has little trend if the ADX moves below both the DI lines.

This system can be used by itself or as a filter to the SAR system.

Finally, there is the *average directional movement index rating (ADXR)*. See Figure 9.8. Wilder created the ADXR as a measuring tool of the strength of the ADX. Calculate the ADXR as the average of the current ADX and the ADX 14 days ago:

$$ADXR = \frac{ADX_C + ADX_{14 \text{ days back}}}{2}$$

Market Profile®

This indicator was designed by J. Peter Steidlmeyer in an attempt to analyze how much trading activity takes place in the futures market, at what price, and at what time. The assumption of the Market Profile® is that price activity forms the bell, or normal distribution, curve about 80% of the time. Figure 9.9 displays a diagram of the bell curve with the percentage of all possible outcomes for each of the standard deviations. One standard deviation covers 68.3% of all outcomes, two standard deviations take in 95.4% of all outcomes, and three standard deviations cover 99.7% of all outcomes.

Caution: To fit more curves on the monitor page, the bell curves are positioned vertically. The wider and lower the bell curve is, the more prices were traded, and vice versa.

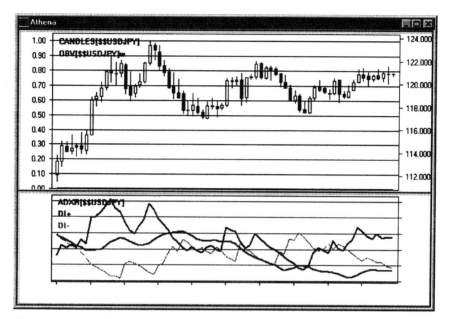

Figure 9.8. Example of ADXR with +DI and –DI on the US dollar/Japanese yen chart. (*Source:* Bridge Information Systems, Inc.)

The indicator divides the trading time into 30-minute vertical stripes. In each stripe you can see several letters. Each letter labels all trades that occurred in each specific 30-minute period. The letters are used in alphabetical order. You start with capital letters and, if the need occurs, continue with lower case. When you have two letters in the same stripe, it means that two different series of prices were traded during two different periods of time. For instance, in the first 30-minute stripe you see A and C. Letter A is marked next to the British pound futures prices 1.6000 to 1.6050, which means that this range is traded in the first half hour of trading, say 8:30 to 9:00. Letter C, above A in the same stripe, is posted next to the range 1.6060 to 1.6100, and shows that this range was traded in the third half hour, or 9:30 to 10:00.

Conceivably, if price activity creates half of the bell curve in the first part of the trading day, the market can be expected to trade on the other side for the balance of the day. In addition, in a normal market, the trading range is established in approximately the first hour.

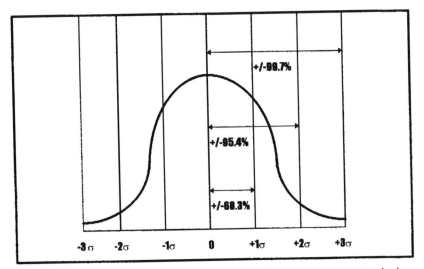

Figure 9.9. The bell (normal distribution) curve, with the outcomes per standard deviation.

Steidlmeyer identified four trading patterns and their respective probabilities of occurrence:

- *Normal days (80%).* Figure 9.10 illustrates the general pattern of a normal trading day.
- *Two types of trending days (15%).* Figure 9.11 displays the two general patterns of trending days. The patterns of trending days may be more elongated due to steadily advancing or declining markets (A), or they may look stocky due to gapping markets in search of new extreme prices (B).
- *Nontrending days (5%).* Figure 9.12 shows the diagram of a nontrending day.

Figure 9.13 provides a market example of the Market Profile® on the IMM Japanese yen futures.

```
            115.20        C
            115.19        C
            115.18        C
            115.17        AC
            115.16        ADG
            115.15        ADF
            115.14        G
            115.13        ADE
            115.12        AE
            115.11        A
            115.10        B
            115.09        B
```

Figure 9.10. A normal trading day.

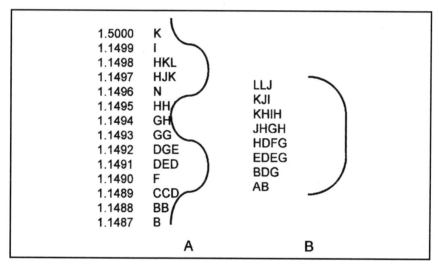

Figure 9.11. The two patterns of the trending days.

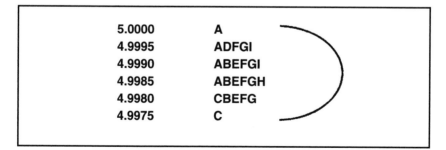

Figure 9.12. The pattern of a nontrending day.

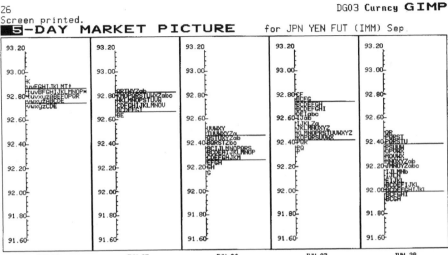

Figure 9.13. The Market Profile® on the IMM Japanese yen futures.
(*Source:* Bloomberg Financial Services.)

On Balance Volume (OBV)

The *on balance volume* is an indicator developed by Joseph Granville for the stock market back in 1963. Since it requires volume information, OBV can be used only in the currency futures market. OBV consists of a line that confirms the soundness of the current trend. See Figures 9.14 and 9.15.

To calculate OBV, use the following formula:

$$CL_{C-1} - CL_C = \Delta$$

where

CL_C = current closing price
CL_{C-1} = previous closing price
Δ = price change

if

$$\Delta > 0$$

you add it to the previous series of numbers.

$$\Delta < 0$$

you deduct it from the previous series of numbers.

The direction of the OBV is more significant than volume itself. In fact, OBV assumes an arbitrary figure as the starting point. Generally, OBV follows the trend. Interpret a divergence as a reversal signal.

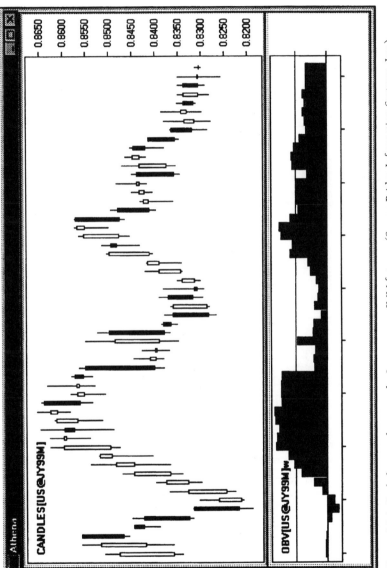

Figure 9.14. On balance volume on the Japanese yen IMM futures. (*Source*: Bridge Information Systems, Inc.)

269

Volume Accumulation Oscillator (VAO)

An alternative to the on balance volume is the *volume accumulation oscillator.* This indicator, designed by Marc Chaikin, is a more sensitive measure of the volume versus price than the OBV. The VAO allocates a proportional amount of volume to the mean price of the period, depending on the position of the closing price versus the mean price. A close above the mean price is assigned a positive value, and a close below that has a negative value. Volume is assigned fully when the currency closes at either the high or the low of the period.

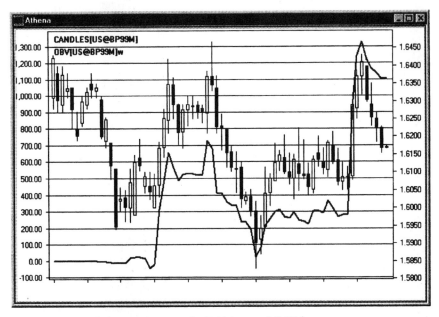

Figure 9.15. On balance volume on the British pound IMM futures. (*Source:* Bridge Information Systems, Inc.)

The calculation is done as follows:

$$\text{Volume Accumulation Oscillator} = \frac{\dfrac{(CL - L) - (H - CL)}{H - L} \times V}{1000}$$

where

CL = closing price
H = high
L = low
V = volume

Sequential Analysis®

Tom DeMark devised an interesting and powerful indicator called the *Sequential®*, which operates on the assumption that price reversals are the result of market exhaustion. He designed this indicator precisely to identify market exhaustions and therefore put you in the driver's seat.

The Sequential® consists of three steps:

1. Setup.
2. Intersection.
3. Countdown.

Setup

The first phase compares a series of nine consecutive closes with the closing price four periods earlier. Most of the time, the completion of this series triggers at least a minor reversal. This comparison yields two types of setup. A *buying setup* occurs when a series of nine consecutive closes are lower than the close four periods earlier (see Figure 9.16). The countdown starts past day 4, where the close of day 4 acts as the close cap. Conversely, a *selling setup* results if a series of nine consecutive closes are higher than the close four periods earlier (see Figure 9.17). There are no exceptions with regard to these parameters.

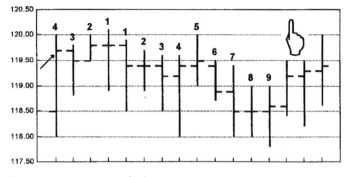

Figure 9.16. Tom DeMark's buying setup.

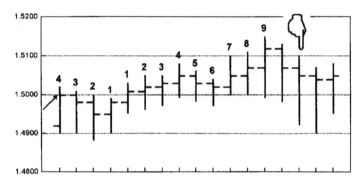

Figure 9.17. Tom DeMark's selling setup.

Intersection

This step was designed to filter out false expectations of a reversal when a trend is overpowering. To this end, Tom DeMark says that the lows of day 8 or 9 must intersect the highs of three or four days earlier in the setup phase to complete the sell setup. Conversely, the highs of day 8 or 9 must intersect the lows of three or four days earlier in the setup phase to confirm the buying setup. The failure of the intersection really means the confirmation of a strong trend.

Countdown

This should be the last phase before reversal. The countdown begins on bar 9 of the setup or when the intersection occurs (whichever comes first).

Buy setup = 13 closes (not necessarily consecutive) ≤ low 2 periods earlier
Sell setup = 13 closes (not necessarily consecutive) ≥ high 2 periods earlier

Generally, when both 9 and 13 counts are complete, the reversal areas are identified and you can act on the trend reversal. Figures 9.18 and 9.19 show you market examples of the sequential indicator.

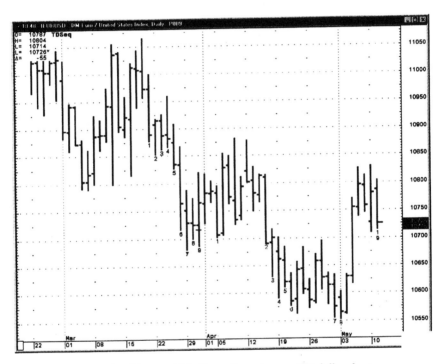

Figure 9.18. Example of Sequential Indicator® on the euro/US dollar chart. (*Source:* CQG. ©Copyright CQG INC.)

Range Expansion Index (REI)™

Tom DeMark developed this indicator to identify more precisely trend patterns and trading signals. He compares the current prices with the prices two days earlier, rather than the customary yesterday's prices, to eliminate insignificant short-term events.

The Range Expansion Index™ is designed to be sensitive only in trending periods; it is lackluster in sideways markets and during steep

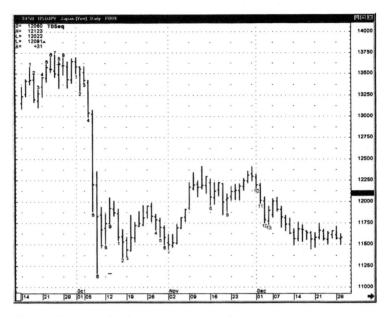

Figure 9.19. Example of Sequential Indicator® on the US dollar/Japanese yen. (*Source:* CQG. ©Copyright CQG INC.)

trading. Furthermore, DeMark presents certain conditions (shown in the following calculations) to prevent premature position taking.

The calculation of the Range Expansion Index™ is done in the following way:

$$D_1 = (H_C - H_{C-2}) + (L_C - L_{C-2})$$

where

D_1 = changes for day 1
H_C = current high
H_{C-2} = high two days before
L_C = current low
L_{C-2} = low two days before

Conditions:

1. $H_C \geq$ low five or six days before.
    or
    $H_{C-2} \geq$ low seven or eight days before.

2. $L_C \leq$ high five or six days before.

or

$L_{C-2} \leq$ high seven or eight days before.

If either of the conditions is not met, the value of the day is zero.

The calculations for D_1 are repeated for the next four days: D_2, D_3, D_4, and D_5.

$$REI = \frac{\sum_{i=1}^{5} D_i}{\sum_{i=1}^{5} |D_i|} \times 100$$

where the denominator is in absolute value (all results have positive values).

REI™ fluctuates on a scale of −100 to +100. The currency is overbought when REI™ exceeds +45 and oversold when REI™ falls below −45. If REI™ spends less than five consecutive periods in the overbought/oversold area, you must expect at least a minor reversal. If it stays above or below these levels for more than five consecutive periods, the signal is muddled. REI™ must then return to the neutral −45/+45 zone, and a reversal signal results only if it stages a second signal outside the neutral zone.

Figure 9.20 shows you a market example of the Range Expansion Index™ in the euro/Japanese yen.

TD Arcs™

Tom DeMark developed the TD Arcs™ to improve retracement analysis by including both the price action and the time frame. To maintain a similar scale on different charts, he anchors the fulcrum of the retracement arc. Figure 9.21 shows you a market example of the TD Arcs™ in the euro/US dollar. For an up side arc, he draws a diagonal line from the recent low to the highest price following the significant low. On this diagonal line, he marks the Fibonacci percentage reversal figures: .382 and .618. Finally, he generates the TD Arc™ by swinging the line forward

Figure 9.20. An example of Range Expansion Index™ on the euro/Japanese yen chart. (*Source:* CQG. ©Copyright CQG INC.)

Figure 9.21. An example of the TD Arcs™ on the euro/US dollar chart. (*Source:* CQG. ©Copyright CQG INC.)

from the fulcrum (which is placed at the low). The process is reversed for a down side arc.

DeMark applies a time filter to measure the strength of the retracement. If the currency retraces .382 percentages of the previous move in less than .382 percentages of the periods that generated the previous move, then the currency has a higher probability of retracing further to .618 percentages.

Triple Exponential Smoothing Oscillator (TRIX)

TRIX was developed by John Hutson to generate trading signals in the stock market. Although less popular in foreign exchange, it may be worth a try. The oscillator is based on triple exponentially smoothed moving averages and their momentum. As it swings on an open scale, TRIX generates a buying signal at an extreme negative level and a selling signal at an extreme positive level. See Figures 9.22 and 9.23.

Figure 9.22. Example of TRIX on the US dollar/Japanese yen chart. (*Source:* CQG. ©Copyright CQG INC.)

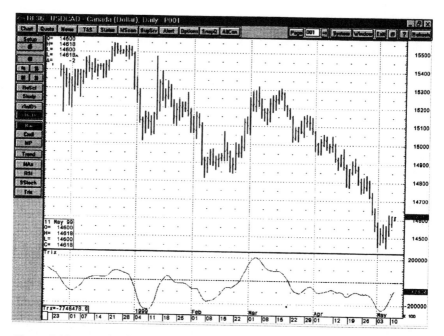

Figure 9.23. Example of TRIX on the US dollar/Canadian dollar chart. (*Source:* CQG. ©Copyright CQG INC.)

TRIX is more helpful with timing rather than with direction. Therefore it should be used along with the parabolic or CCI.

Swing Index

The *Swing Index* was designed by Welles Wilder to ascertain the direction of the currency and the potential changes in this direction, based on the price differences between the current period and the previous one.

This index fluctuates on a scale of −100 to +100, and its extreme levels suggest reversal.

To calculate the Swing Index, you apply the following formula:

$$SI = \frac{CCL - PCL + \dfrac{CCL - COP}{2} + \dfrac{PCL - POP}{4}}{TR} \times \frac{\text{Max}}{L}$$

where

CCL = current closing
PCL = previous closing
COP = current opening
POP = previous opening
TR   = true range
Max = the greater of the difference between the current high and the previous close or the difference between the current low and the previous close
L   = directional limit move

Accumulation Swing Index (ASI)

The *Accumulation Swing Index (ASI)* is based on the Swing Index. A *buying signal* is generated when the daily high exceeds the previous SI significant high, and a *selling signal* occurs when the daily low dips under the significant SI low.

Figures 9.24 and 9.25 illustrate the Accumulation Swing Index in the currency market.

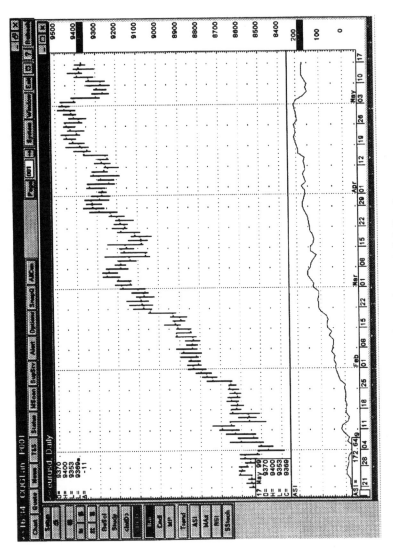

Figure 9.24. The Accumulation Swing Index on the euro/US dollar chart.
(*Source:* CQG. ©Copyright CQG INC.)

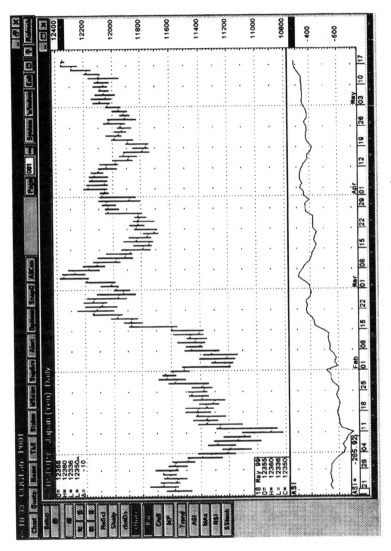

Figure 9.25. Accumulation Swing Index on the US dollar/Japanese yen chart.
(*Source:* CQG. ©Copyright CQG INC.)

W.D. GANN ANALYSIS

One of the most influential methods of technical analysis was developed by William D. Gann (1878–1955), the eminent stock and commodity trader. The complex Gann analysis is based on traditional chart formations but his mathematical approach generates unique trading signals.

The most important features of his analysis are:

1. Percentage retracements.
2. Geometric angles.
3. The cardinal square.
4. Geometric forms.
5. Squaring of price and time.

Due to the wonders of the electronic charting services, retracement percentages and geometric angles are probably the most used tools from Gann's rich technical analysis closet. They are all just a couple of mouse clicks away.

Percentage Retracements

Gann divided price movements into eighths and thirds, the same percentages he used to calculate the price retracements. The resulting percentages are therefore as follows:

1/8	2/8	1/3	3/8	4/8	5/8	2/3	6/8	7/8	8/8
12.5%	25%	33%	37.5%	50%	62.5%	67%	75%	87.5%	100%

The most important retracement percentage is 50%, and each of the other numbers' importance decreases symmetrically on the upper and lower side. The most significant percentages, used as retracement levels, are in bold.

The 33%, 50%, and 67% levels fully coincide with the retracement percentages from the Dow Theory and roughly to the Fibonacci retracement ratios of 37.5% and 62.5% levels. Figure 10.1 illustrates Gann's retracement percentages.

Geometric Angles

Geometric angles are trend lines drawn from significant highs and lows at specific angles (see Figures 10.2 and 10.3). As mentioned in Chapter 1, Gann considered the 45-degree angle as the most important one. It is the most sustainable rate of price change in the medium or long term. A sharp rally is destined to flatten out faster because it is impossible to sustain the high velocity. Conversely, a lower geometric angle suggests that the supply is too heavy and that it is only a matter of time before the bulls will call it quits and join the bears. The penetration of the 45-degree line signals a trend reversal.

On charts, the 45-degree line is represented as 1×1. Steeper trend lines are determined by a ratio of 1×8, 1×4, 1×3, and 1×2. Flatter trend lines are determined by price and time ratios of 2×1, 3×1, 4×1, and 8×1. Gann also included 1×3 and 3×1 lines, which he felt are more useful for the weekly and monthly charts. This application, known as speedlines, is currently used on its own.

The labeling is very useful because you cannot accurately measure angles on electronic charts. As you squeeze in more prices, the angle becomes more and more distorted. So, rather than wasting your time in futile contortionist positions, proceed with your analysis simply by reading the labels.

1×8	1×4	1×3	1×2	1×1	2×1	3×1	4×1	8×1
$82^{1}/_{2}°$	$75°$	$71^{1}/_{4}°$	$63^{3}/_{4}°$	$45°$	$26^{1}/_{4}°$	$18^{3}/_{4}°$	$15°$	$7^{1}/_{2}°$

These lines work in the same manner as speedlines. The penetration of one line suggests that the currency will trade between the next two

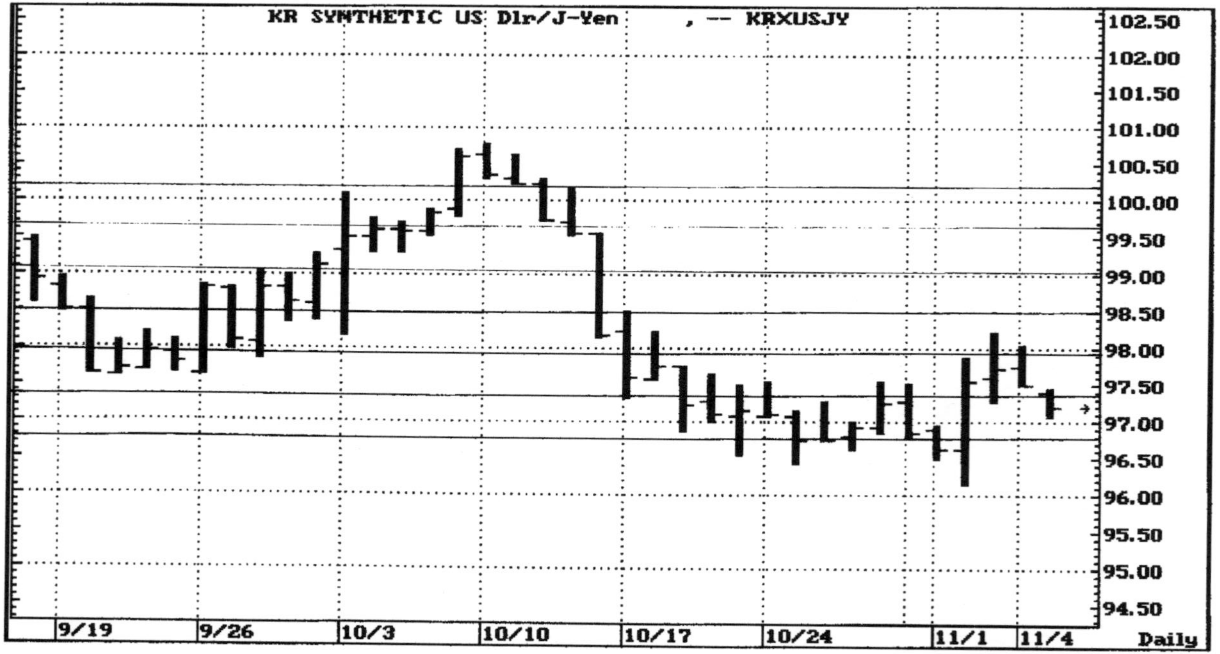

Figure 10.1. Gann's retracement percentages in a US dollar/Japanese yen daily bar chart. (*Source:* Bridge Information Systems, Inc.)

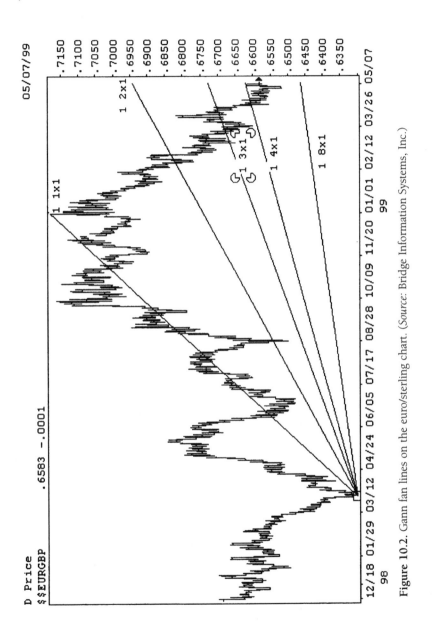

Figure 10.2. Gann fan lines on the euro/sterling chart. (*Source:* Bridge Information Systems, Inc.)

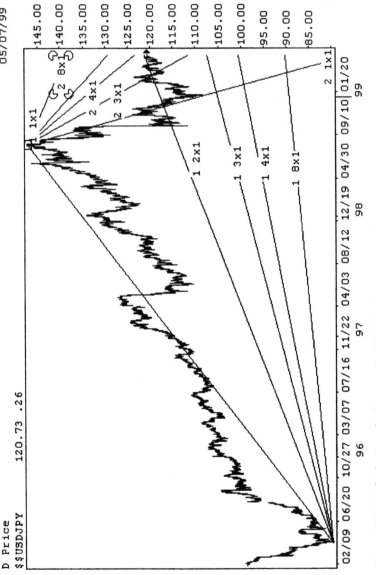

Figure 10.3. Multiple Gann fan lines on the US dollar/Japanese yen chart. (*Source:* Bridge Information Systems, Inc.)

lines. The intersection of two sets of geometric angles at a 90-degree angle generates an even more significant level.

The 33% and 67% speedlines tend to work better on weekly or monthly charts.

The Cardinal Square

The *cardinal square* is a unique technique of forecasting significant chart points by counting from the all-time low price of the currency. It consists of a square divided by a cross into four quadrants. The all-time low price is housed in the center of the cardinal cross. The following higher prices are entered in a clockwise direction. The rates positioned in the cardinal cross are the most significant chart points, both for support and resistance. Figure 10.4 shows an application for the sterling/dollar, and Figure 10.5 provides the application for the US dollar/Japanese yen based on their all-time lows reached in the spring of 1995. The lowest prices— 1.0345 and 79.75, respectively—are entered at the center of the cross. See also Figure 10.6.

The cardinal square is a powerful mechanical tool. There are no calculations to make or trend lines to draw. The table remains unchanged for as long as the low in the currency stays the same. Generally, these prices are significant and hold at least during the initial attacks. When broken, these levels trigger 50-pip moves. Naturally, you cannot expect all the rallies or selloffs through these "magic" numbers to be limited to 50 points. At times, very aggressive markets exceed these targets, whether trading or gapping. Although the cardinal square is remarkable for its consistency and its low rate of failure, traders must adjust their arsenal of weapons to the style of fight.

Geometric Forms

In his analysis, Gann used basic forms of geometry, such as the square, the circle, and the triangle. The 360 degrees are a staple in his time analysis. To reach his time targets, he counted forward from significant chart points by 30, 90, 120, 180, and 360. These forward days are potential reversal days. See Figure 10.7.

Figure 10.4. The Gann cardinal square for the sterling/dollar. (*Source:* ©Copyright 1999 Cornelius Luca.)

Figure 10.5. The Gann cardinal square for the US dollar/Japanese yen. (*Source:* ©Copyright 1995 Cornelius Luca.)

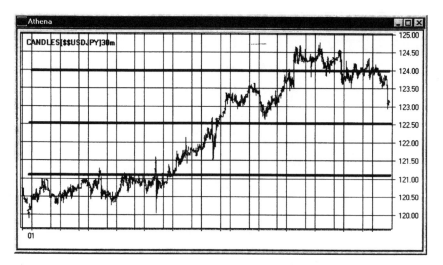

Figure 10.6. Application of Gann 50-pip pivot points on the US dollar/Japanese yen chart. (*Source:* Bridge Information Systems, Inc.)

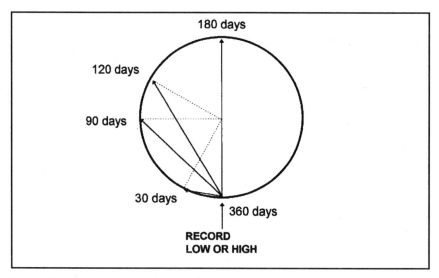

Figure 10.7. Gann's concept of potential reversal days based on circle and circle arches.

Gann also considered the number 7 as a technically significant number.

The Squaring of Price and Time

The squaring of price and time is Gann's technique of converting a significant commodity dollar price into time units (from days to years) and adding these time periods to the day on which the significant price was reached. When the time targets are reached, time and price are squared, and the market is likely to reverse.

For instance, the record low of the US dollar/Japanese yen was reached on April 19, 1995 at 79.75, or, rounded up, 80.00. Now you count and extrapolate 80 days, 80 weeks, and 80 months from this date. These dates are all potential reversal dates. The 80-day target was July 8, 1995. The 80-week target was Wednesday, October 15, 1996. Finally, the 80-month target is December 19, 2002. When the target day falls on a holiday or weekend, such as the US dollar/Japanese yen 80-day target, which fell on a Saturday, look for either the previous day (Friday, July 7, 1995), or the next business day (Monday, July 10, 1995).

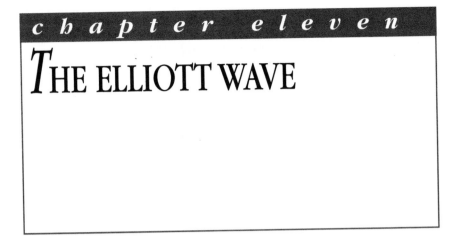

chapter eleven

THE ELLIOTT WAVE

Ralph Nelson Elliott developed his theory on technical analysis in 1938, during a long period of convalescence. Trained as an accountant, Elliott studied technical analysis under the influence of the Dow Theory. He concluded that the financial markets in general and the stock market in particular have a striking resemblance to a basic harmony found in nature. In his article "The Wave Principle," Elliott postulated that price movements in financial markets are repetitive in form but not necessarily in time or amplitude.

Basically, the Elliott Wave principle is a system of empirically derived rules for interpreting action in the markets. It is a unique tool both in terms of value and in terms of difficulty of use. The method can be applied to any of the financial markets, and its accuracy in identifying direction changes in advance can be remarkable. Yet many traders are intimidated by the intricacies of the Elliott Wave. Accurately reading and measuring the waves can be a daunting task. Perhaps nobody has worked harder than Robert Prechter to bring Elliott's theory to traders in a clear and comprehensive manner. He has customized this method to all major financial markets and, along with A. J. Frost, he wrote the definitive book on the subject, *Elliott Wave Principle*.

The Elliott Wave principle is a comprehensive analysis that covers all identifiable formations in the financial markets. This analysis is helpful for both full-time Elliott students and for seasoned analysts. If you only remember, for instance, the failure of wave 5, then you can make a hefty profit simply by entering a position against the direction of this wave. Naturally, the more versed you are in Elliott's intricacies, the more opportunities for profit you will have. And if you are somewhat turned off by all those tricky combinations and subdivisions, don't despair. Help is available. Despite the complexity of Elliott Wave analysis, software programs can do the counting for you. All you have to do is place the trade.

Basics of Wave Analysis

In a series of articles published in 1939 by *Financial World* magazine, Elliott held that a financial market unfolds according to a basic rhythm or pattern: five waves in the direction of the trend at one larger scale and three waves against that trend. See Figure 11.1. In a rising market, this five-wave/three-wave pattern forms a complete bull market/bear market cycle of eight waves. The five-wave upward movement as a whole is referred to as an *impulse* wave, while the three-wave countertrend movement is described as a *corrective* wave.

Within the five-wave bull move, waves 1, 3, and 5 are themselves impulse waves. They are subdivided into five waves of smaller scale. The subwaves of impulse sequences are labeled with numbers. Waves 2 and 4 are corrective waves, subdividing into three smaller waves each. The subwaves of corrections are labeled with letters (see Figure 11.1).

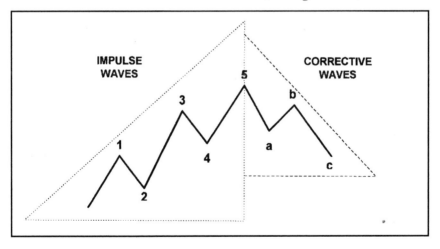

Figure 11.1. The basic Elliott Wave pattern.

Following the cycle shown in Figure 11.2, a second five-wave up side movement begins, followed by another three-wave correction, followed by one more five-wave upmove. This sequence of movements constitutes a five-wave impulse pattern within the larger trend, and a three-wave corrective movement at the same scale must follow.

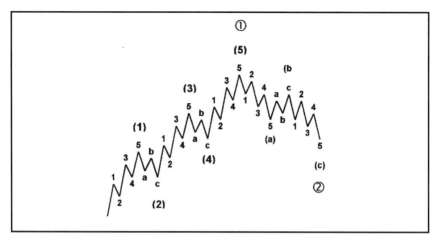

Figure 11.2. The larger scale pattern in detail.

As you can see in Figure 11.2, waves of any degree in any series can be subdivided and then subdivided again into waves of a smaller degree, or they can be grouped into waves of a larger degree. This structure is an example of "fractal" geometry, incorporating self-similarity and scaling symmetry, and is consistent with numerous scientific insights into the nature of growth and evolution.

The general impulse and corrective waves, which together form a cycle, are sorted and classified in the following table:

	Impulse	*Corrective*	*Cycle*
Waves	1	1	2
First subdivisions	5	3	8
Second subdivisions	21	13	34
Third subdivisions	89	55	144

The form of market movement is essentially the same, regardless of the duration or size of the currency movements. Smaller-scale movements link up to create larger-scale movements possessing the same basic form. Conversely, large-scale movements consist of smaller-scale subdivisions with which they share a geometric similarity. Because these movements link up in increments of five waves and three waves, they generate sequences of numbers that the analyst can use along with the rules of wave formation to identify the current state of pattern development, as shown in Figure 11.3.

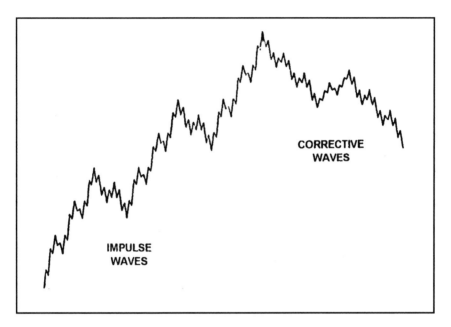

Figure 11.3. A complete market cycle.

Elliott identified three essential rules of interpretation of his wave principle:

1. A second wave may never retrace more than 100% of a first wave. As you can see in Figure 11.4, for example, the beginning of the first wave in a bull British pound market is at 1.5000. The low of the second wave may not go below 1.5000. Since the second wave reached a low of 1.4900, the whole reading is canceled.
2. In an impulse sequence, the third wave is never the shortest wave; often it is the longest. See Figure 11.5.
3. With the minor exception of one specific type of wave pattern, wave 4 can never enter the price range of wave 1. As illustrated in Figure 11.6, if the top of wave 1 in a bull USD/JPY market is 115.00 and if wave 4 falls below this price level reaching a low of 114.00, then the count is stopped.

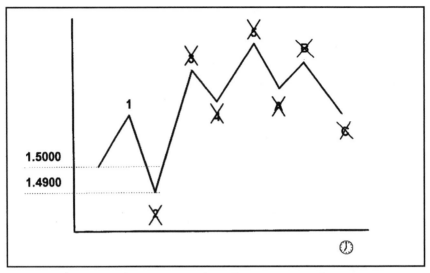

Figure 11.4. First rule of interpretation of the Elliott Wave Principle.

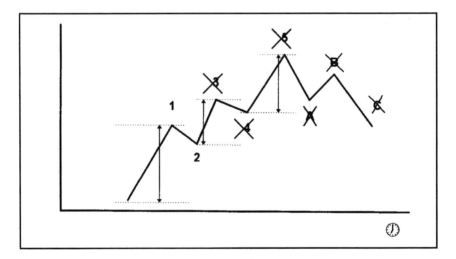

Figure 11.5. Second rule of interpretation of the Elliott Wave Principle.

Characteristics of the Waves

Robert Prechter, independently of Elliott's research, identified specific characteristics of each wave. This contribution has proved to be helpful for "wave hunters."

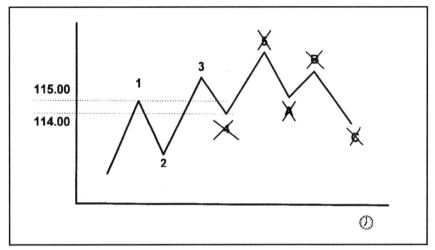

Figure 11.6. Third rule of interpretation of the Elliott Wave Principle.

Wave 1. Wave 1 is difficult to identify since it appears to be more of a correction. It may be dynamic, as when the price breaks out of a consolidation, and then again it may not. It is often the shortest of the impulse waves.

Wave 2. Wave 2 should be easier to identify due to its three-subwave structure. It tends to retrace by about .618% (a Fibonacci percentage retracement) of the first wave. Of course, if it retraces more than 100%, then the reading is off. This wave tends to generate a major reversal formation.

Wave 3. Wave 3 tends to be the longest and, as shown in Figure 11.4, it is never the shortest. It has a dynamic move, and the penetration of the high of wave 1 attracts more buyers in a bull market or more sellers in a bear market. Naturally, this makes for a good volume. Fundamentals tend to support the move and the psychology of the market exudes optimism.

Wave 4. According to the *rule of alternation,* if wave 2 is complex, then wave 4 tends to have a less complex pattern, and vice versa. This wave is prone to developing a triangular formation. You also remember from Figure 11.5 that wave 4 cannot overlap wave 1.

Wave 5. Wave 5 can be dynamic and extended in foreign exchange. By now everyone has figured out the long-term trend, the fundamentals are packaged in a rosy wrap, and a good time is being had by all. These overall conditions can create a good overshooting scenario. Oscillators should help.

Wave A. Wave A is difficult to catch through all the euphoria. A good hint comes from the break into five subwaves.

Wave B. Wave B may be of different complexities and lengths, since the last bulls are making their final mark in a previously rising market and the bears are testing the waters and starting to get short currency.

Wave C. Wave C puts the stamp on the end of the trend. Following a bull market, wave C should fall below the bottom of wave A.

Impulse Waves—Variations

Extensions

In the five-wave sequence, one of the three impulse subwaves (i.e., wave 1, wave 3, or wave 5) tends to generate an *extension*—an elongated movement, usually with internal subdivisions. At times, these subdivisions are of nearly the same amplitude and duration as the larger-degree waves of the main impulse sequence, giving a total count of nine waves of similar size rather than the normal count of five for the main sequence. In a nine-wave sequence, identifying which wave is extended is sometimes difficult. This is usually irrelevant, though, because nine and five counts have a similar technical significance. Figures 11.7 and 11.8 illustrate this point, since examples of extensions in various wave positions make it clear that the overall significance is the same in each case.

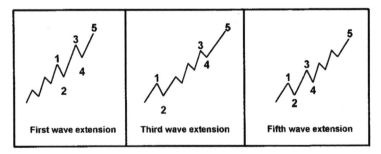

Figure 11.7. Wave extensions in a bull market.

Extensions can be useful guides to the lengths of future waves. Most impulse sequences contain extensions in only one of their three impulsive

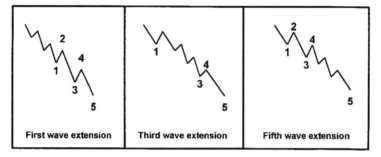

First wave extension · Third wave extension · Fifth wave extension

Figure 11.8. Wave extensions in a bear market.

subwaves. Therefore, if the first and third waves have about the same magnitude, the fifth wave will probably be extended, especially if volume during the fifth wave is greater than during the third. Conversely, if wave 3 has already extended itself, wave 5 should be simple and resemble wave 1.

Extensions may also occur within extensions. Although extended fifth waves are not uncommon, extensions of extensions occur most often within third waves.

Diagonal Triangles

Elliott identified two types of diagonal triangle. Both types occur relatively rarely, and each has highly specific implications for future market movements. These formations are different from the common corrective triangles found in fourth wave positions within impulse waves and B wave positions within corrections.

The first type of diagonal triangle occurs only in fifth waves and C waves, and it signals that the preceding move was too volatile. Essentially a rising wedge formation defined by two converging trend lines, a type 1 diagonal triangle indicates exhaustion of the larger movement. Unlike other impulse waves, all the pattern's subwaves, including waves 1, 3, and 5, consist of three-wave movements, and their fourth waves often enter the price range of their first waves.

A rising diagonal triangle type 1 is bearish, since it is usually followed by a sharp decline, at least to the level where the formation began. Conversely, a falling diagonal triangle type 1 is bullish, since an upward thrust usually follows.

The second type of diagonal triangle occurs even less frequently than the first type. This pattern, found in first wave or A wave positions in very rare cases, resembles a diagonal type 1 in that it is defined by converging trend lines and its first and fourth waves overlap. It differs significantly from type 1 because its impulsive subwaves (waves 1, 3, and 5) are normal, five-wave impulse waves, versus the three-wave subwaves of type 1. This is consistent with the message of the type 2 diagonal triangle, which signals the continuation of the underlying trend, in contrast to the type 1's message of its termination.

Failures or Truncated Fifths

According to Elliott, any impulse pattern in which the extreme of the fifth wave fails to exceed the extreme of the third wave was a *failure*. Figure 11.9 shows failure in a bull market in USD/CHF. The top of the third wave reaches 1.2100, but the fifth wave stops at 1.2000, failing to exceed 1.2100. Consequently, the market stages a sharp selloff in USD/CHF.

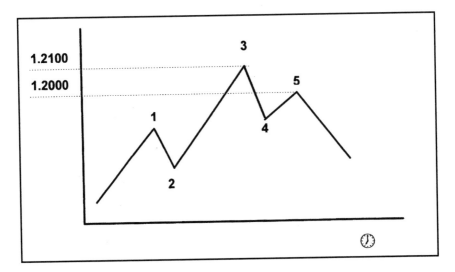

Figure 11.9. Bull market failure.

Conversely, Figure 11.10 illustrates a fifth-wave bear market failure. Wave 3 reaches a low of 130 in a bearish USD/JPY market. The truncated fifth wave is generated when it only reaches a low of 131.00, thus triggering a sharp rally in USD/JPY.

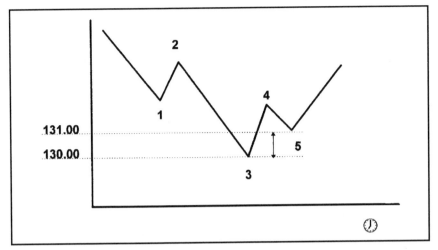

Figure 11.10. Bear market failure.

As the illustrations show, the truncated fifth wave contains the necessary impulsive substructure to complete the larger movement. However, its failure to surpass the previous impulse wave's extreme signals weakness in the underlying trend, and a sharp reversal usually follows.

Corrective Waves

Naturally, currency swings tend to move more easily *with* the trend of a larger degree than *against* it. Therefore, corrective waves can be highly complex, choppy, and often difficult to interpret before completion. Since the terminations of corrective waves are less predictable than those of impulse waves, traders must exercise greater caution when the currency is in a corrective mood rather than when the trend is clearly impulsive. Moreover, while only three main types of impulse wave exist, there are a total of ten basic corrective wave patterns, which can form extended corrections of great complexity.

The single most important characteristic about corrections is that they never consist of five subwaves. Only impulse waves consist of five subwaves. An initial five-wave movement against the larger trend can never be a complete correction, but only part of it.

Elliott identified the following four corrective patterns:

1. Zigzags (5-3-5).
2. Flats (3-3-5).
3. Triangles (3-3-3-3-3).
4. Combined structures.

Zigzags

Zigzags are simple three-wave patterns, subdivided into 5-3-5 structures, in which the extreme of wave B remains a significant distance from the beginning of wave A. On rare occasions, a double zigzag formation may occur. Figure 11.11 shows examples of zigzags in a bull market, and Figure 11.12 illustrates zigzags in a bear market.

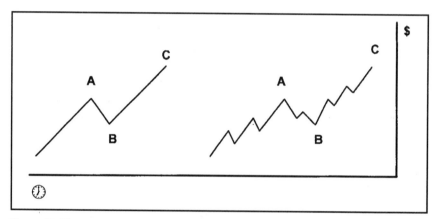

Figure 11.11. Zigzag corrections in a bull market.

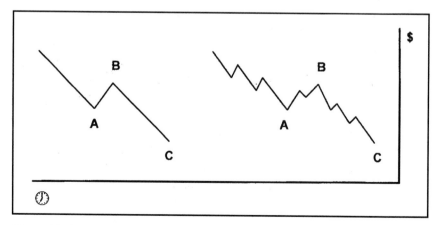

Figure 11.12. Zigzag corrections in a bear market.

Flats

Flat corrections have 3-3-5 structures. The original movement of wave A lacks the momentum to develop into full five waves, as in a zigzag. Wave B also lacks countertrend pressure and often ends at or beyond the start of wave A. Finally, wave C usually finishes near the extreme of wave A, rather than significantly beyond it, as in a zigzag.

Flat corrections, overall, are less damaging than zigzags to the current currency trend. They tend to indicate strength underlying the larger trend and thus often precede or follow extensions. As a rule, the longer the flat, the more dynamic the next impulse wave is.

Elliott identified four types of flats: regular, expanded, irregular, and running.

1. *Regular flats* occur when wave B ends at or just beyond the level of the *beginning* of wave A and wave C ends at or slightly beyond the level of the *extreme* of wave A. See Figure 11.13.

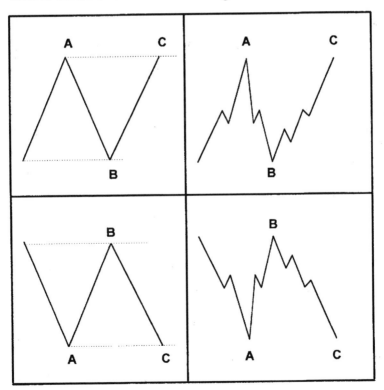

Figure 11.13. Regular flat corrections.

2. *Expanded flat* corrections can be identified when wave B significantly exceeds the level of the start of wave A, and wave C significantly exceeds the level of the extreme of wave A. See Figure 11.14.

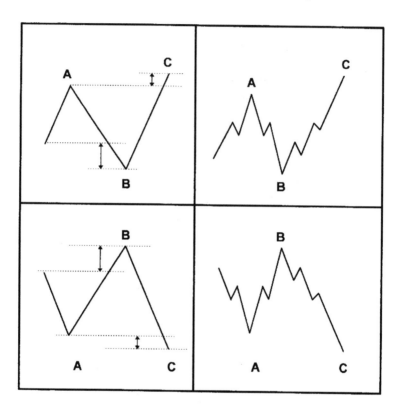

Figure 11.14. Expanded flat corrections.

3. *Irregular flats* can be identified by the fact that, while wave B ends near the start of wave A, similar to a regular flat, wave C fails to move all the way to the extreme of wave A. See Figure 11.15.

4. *Running flats* are a rare type of formation in which wave B greatly exceeds the beginning of wave A, while wave C fails to exceed that level.

Flats reflect the strength of the significant trend and occur when the market is so volatile that corrective patterns have no time to develop normally.

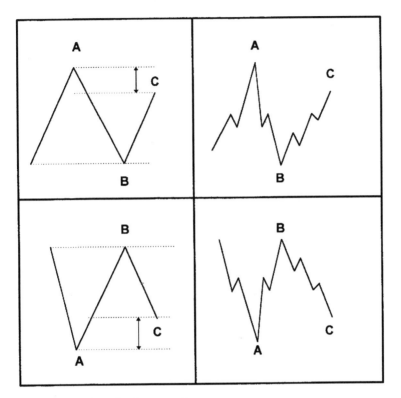

Figure 11.15. Irregular flat corrections.

Triangles

Triangles tend to occur just before the final currency attack in the direction of the long-term trend. They tend to be prolonged because of low volume and low volatility during a consolidation period.

Triangles consist of five waves, labeled A-B-C-D-E, subdivided into three waves each. The four types of triangles discussed in Chapter 3 (symmetrical, ascending, descending, and expanding) are illustrated in Figure 11.16.

The trend lines that contain a triangle are usually highly accurate. The only exception may be wave E, which often overshoots the trend line, especially in expanding and contracting triangles.

After completion of a triangle, the final impulse wave of the larger trend is usually swift and has a price objective equal to the base of the triangle. This movement is called a *thrust*.

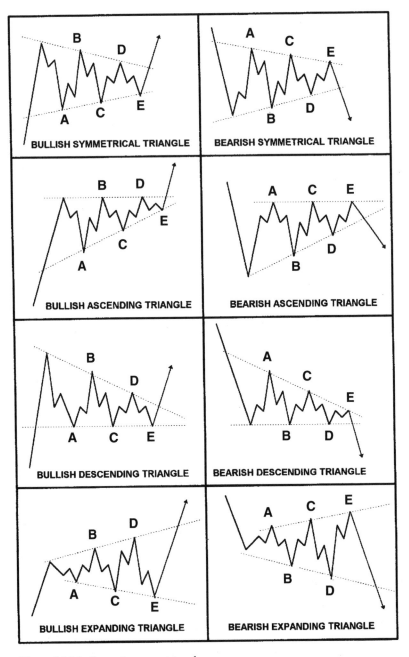

Figure 11.16. Corrective wave triangles.

Combined Structures

In Elliott Wave analysis, zigzags and flats are often referred to as *threes*. At times, these patterns, along with triangles, may combine into more complex *double three* or *triple three* formations. These combined structures consist of two or more threes separated by smaller three-wave movements, labeled X waves. For example, a double three may consist of a flat, a small zigzag-forming wave X, and a second flat, or it might contain a zigzag, a small flat in wave X, and a second zigzag.

The combined structures are generally sideways formations reflecting market hesitation, as the currencies wait for the release of the economic indicators to catch up with market expectations. Elliott also indicated that the combined structures may turn against the trend. The breakout from a double or triple three tends to be forceful.

Figures 11.17 and 11.18 show you applications of the Elliott Wave principle in the currency market.

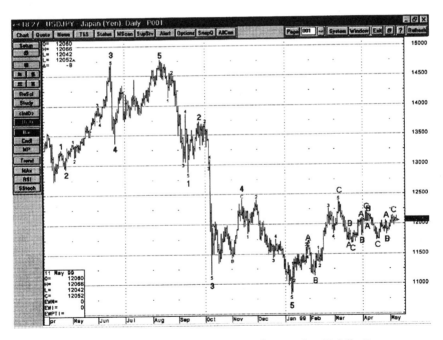

Figure 11.17. An application of the Elliott Wave analysis on the US dollar/Japanese yen chart. (*Source:* CQG. ©Copyright CQG INC.)

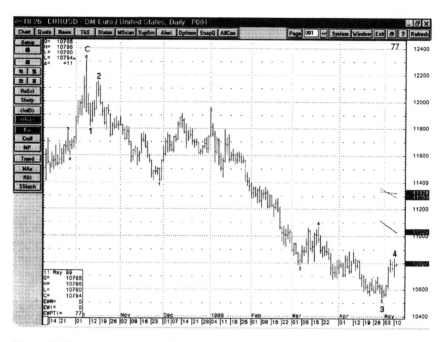

Figure 11.18. An application of Elliott Wave analysis on euro/US dollar. (*Source:* CQG. ©Copyright CQG INC.)

Fibonacci Analysis

One of the most important contributions Ralph Nelson Elliott brought to technical analysis is his discovery that movements of the same degree tend to be related to one another by a specific mathematical ratio, called the *Fibonacci ratio,* or the *golden ratio.*

The Fibonacci ratio is named after Leonardo Fibonacci of Pisa, an Italian mathematician, who introduced Hindu-Arabic numerals to Western Europe in his work titled *Liber Abaci,* or *Book of Calculations,* about seven centuries ago.

The Fibonacci sequence is an ascending series of numbers in which each consists of the sum of the two preceding numbers, starting with 0. The series thus calculated yields the following series of numbers:

1, 1, 2, 3, 5, 8, 13, 21, 34, 55, 89, 144, 233, 377, 610, 987, 1597, 2584, 4181, (etc.)

The Fibonacci series displays the following numerical relationships:

1. The sum of any two consecutive numbers equals the next consecutive number.
2. After the first four numbers, the ratio between two consecutive ascending numbers approaches .618.
3. As the numbers increase, the ratio between two consecutive descending numbers approaches 1.618, which is the inverse of .618.
4. The ratio between two alternate ascending numbers approaches .382, and the ratio between two consecutive descending numbers approaches 2.618.

The ratio .618, known as the Fibonacci ratio, an irrational number, is calculated as follows:

$$\frac{\sqrt{5}-1}{2} \ = \ 0.6180339887499\ldots$$

Since 3 and 5 are Fibonacci numbers (i.e., numbers in the basic Fibonacci sequence), combinations of threes and fives tend to add up to larger Fibonacci numbers. Elliott's basic 5-3-5-3-5 impulse wave sequence adds up to 21, which is also a Fibonacci number. Adding a 5-3-5 corrective wave sequence raises the total to 34, a Fibonacci number. Fibonacci numbers and ratios constitute the basis of the Elliott Wave principle.

The Fibonacci ratio underlies the geometry of the logarithmic spiral, a geometric form found widely in nature. This figure also possesses the noteworthy property that the ratio of the length of the arc to its diameter is 1.618, the inverse of .618.

Many other natural forms exhibit Fibonacci proportions. For example, the double helix of the DNA molecule is characterized by Fibonacci geometry.

The financial markets display Fibonacci proportions in the following ways:

1. Most five-wave sequences contain an extension in one impulse wave.

2. The two nonextended waves tend to be approximately equal in magnitude, and the extended wave tends to be a Fibonacci multiple (most often 1.618 times) of the nonextended waves.

3. Corrective waves tend to retrace a Fibonacci percentage (most often 61.8%, 38.2%, or 23.6%) of the preceding impulse wave.

4. Within corrective waves, Fibonacci relationships often exist among the subwaves.

These relationships constitute a useful trading tool for calculating price targets and placing stops. For instance, if you expect a corrective wave to retrace 61.8% of the preceding impulse wave, you may place a stop slightly below that level. You avoid exposure to excessive loss if the correction is greater than expected.

NOW WHAT?

We are just about at the end of our charting journey. We have trekked from the fundamental trends, through chart formations, through formations specific to bar, candlestick, and point and figure, and through all types of quantitative methods. You now have all the technical tools necessary to succeed in the fast and complex world of foreign exchange. You will not (and should not) use every single tool. If you do, you will only become confused and waste too much time away from trading. However, you must decide for yourself what is good for you, what really helps you make money. Since none of the methods is foolproof, selecting the best mix may be a challenge. This last chapter puts the book's topics into perspective and provides general guidelines.

Most of you are focusing on either the spot or the futures market; so your time horizon is short. However, the best approach is to start your analysis from the large, long-term picture, and narrow it down, step-by-step, to daily or intraday signals (see Figure 12.1). The longer the time frame of a trend or formation, the more technically significant it becomes. Granted, major trends or monthly trend reversal formations do not occur very often. Yet, when they materialize, your current trading should generate a significant profit. On a small scale within a major trend, you should encounter several minor trends. On a grander scale, the triple top formation in GBP/USD shown in Figure 12.2 helped George Soros reap over $1 billion from the Bank of England in just a couple of days.

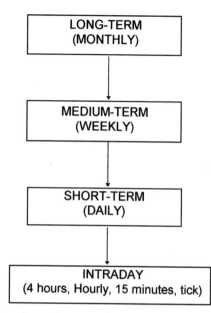

Figure 12.1. Narrowing down a long-term formation to current trading significance.

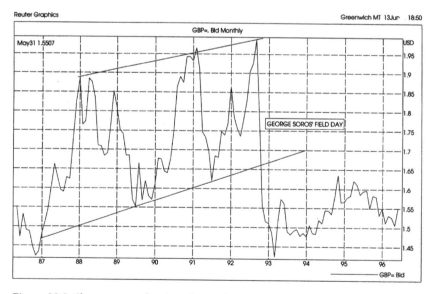

Figure 12.2. Short-term profit taking from a long-term major trend reversal formation in the British pound chart. (*Source:* Reuters.)

The most significant technical tools are the most basic ones. Major trends and their major formations are all that a trader really needs. Once you properly identify them, do not question them and do not hesitate. Just go ahead and trade. The more refined methods are generally of marginal significance. Use those that you find to be of specific help in monitoring either the health of the trend or its propensity to become exhausted. But do not forget that they weigh less on the technical forecasting scale.

Technical Tools for Each Time Frame

Since currency traders have such a diversity of interests—speculation or hedging, short-term trader or long-term trader, interbank trader or proprietary trader—they should use different forecasting methods.

Combining the concepts of time frame and the significance of the analytical tools, chart choices can be categorized as in Figure 12.3.

Long-Term Analysis

When looking at the big picture, you get the clearest view and the most useful signals from line charts and point and figure charts. Your long-term view should not be cluttered by daily highs and lows, which are characteristic of the bar charts and the candlestick charts. Moreover, you must be aware of two potential problems generated by the limited databases of the electronic services.

The first problem is that you may retrieve prices that go back only about ten years. Consider that the overall record high of the US dollar following the demise of the Bretton Woods Accord was registered in 1985 and that its record low occurred in 1995 (the British pound was the exception). You still have access to the record lows, but not to the record highs. Even then, the oldest data might not be fully available; perhaps, for example, only weekly or monthly closings might be available. Thus, the line chart is best suited for the long term.

Why should you care about such old information? You may find cycle analysis helpful to your trading. While it may not be specific enough for the spot trader rushing through one hundred transactions a day, corporate and international fund hedgers may find the old information very helpful. Whereas most of the time this data is of marginal ana-

Long-Term	Medium-Term	Short-Term	Intraday
Line chart	Bar chart	Bar chart	Tick chart
Point and figure chart	Candlestick chart	Candlestick chart	Point and figure chart
Major trend line	Point and figure chart	Point and figure chart	Bar chart
Channel line	Line chart	Minor trend line	Candlestick chart
Gann angles	Secondary trend line	Channel line	Gann's Cardinal Square
Speedlines	Channel line	Classic trend continuation/reversal formations (small)	Stochastics OB/OS and crossovers
Elliott Wave	Classic trend continuation/reversal formations (large)	Support/resistance lines	Momentum (other oscillators)
Gann squaring of price and time	Speedlines	Speedlines	Gaps
	Gann angles	Gann's Cardinal Square	Percentage retracements
	Moving averages	Oscillators	Parabolic
	Currency/oscillator divergence	OB/OS and crossovers	Market Profile®
	TD	Percentage retracements	
	DMI	Parabolic	
	Parabolic		

Figure 12.3. Breakdown of analytical tools according to time frame.

lytical significance, at times it is worth millions. Remember the USD/JPY long-term pattern of 2,000 pips in Chapter 1? A chart going back far enough provides not only the necessary knowledge, but also the confidence to go long USD/JPY 100 million at the low 80s or just above the 100 mark. That time was in the spring and summer of 1995. The next time will be when the USD/JPY will test either the 100 support line or the 120 resistance line.

The second potential problem concerns the use of the point and figure chart. Plotting the point and figure chart requires all prices, or the tick chart. Unfortunately, tick charts only go back several days, generally less than one month.

Both problems can be solved by subscribing to printed chart services, periodically saving charts in your analytical scrap book and even drawing your own.

Once you identify the major trend, your most profitable signals come from the trend and channel lines, as well as from large trend continuation and reversal formations. You receive additional information regarding the health of the trend and the new range from the Gann angles and the speedlines.

Elliott Wave analysis is best suited for long-term analysis for most traders; at the very least, it gives you confidence or enhances it in major or secondary trends. At the very best, you are well enough versed in the Wave's intricacies to apply it in the short term as well.

Only a limited number of traders are likely to use Gann squaring of price and time, but it requires only a small amount of your time to pinpoint the dates and then mark them in your agenda. When the time comes, just keep in mind that the market may reverse.

Medium-Term Analysis

In the medium term, you look at fewer price entries. So basically you can use all the types of charts. Your use of the line chart may be more limited to confirmations of significant breakouts.

Trend lines and channel lines provide you with the most pertinent information, followed by Gann angles and speedlines. In addition, you can follow the signals generated by trend continuation and reversal formations.

You can start to use moving averages for the medium term, especially to hunt for divergences between the underlying currency and its oscillators. Remember that these divergences may last for weeks without any relief.

Tom DeMark's studies should become increasingly popular as more services post them. In general, these studies attempt to determine exhaustion areas and require weeks of data. You may also consider the DMI to gauge the degree of trend presence in a specific currency. The parabolic may also provide some confidence regarding significant directions.

Short-Term Analysis

In the short term, you generally use any of the types of charts, except the line chart. You receive trading signals from the minor trend and channel lines, pertinent support and resistance lines, and small trend formations. You may get signals from the speedlines as the currency approaches them. Gann's Cardinal Square starts becoming very useful in spotting the most significant support and resistance levels.

Oscillators start helping out with either overbought/oversold areas, or with the crossovers of the zero line or among themselves. Directional moves will be balanced by retracements, and you will check the classical 33%, 50%, and 66% as possible retracement targets. The parabolic becomes more significant in the short term, especially in a trending market.

Intraday Analysis

The last step in your analysis, but the most popular step among currency traders around the world, is the intraday trading. For this, you can use all types of charts. As the line chart, you are likely to use the tick chart. Its significance is diminished by its increased sensitivity, but the tick chart will enable you many times to identify several profitable moves even in sideways markets.

Gann's Cardinal Square provides a continuous flow of significant levels. You do not get trend reversal signals, but you receive the most significant support and resistance levels that are not visible on the chart.

Slow and fast stochastics on short-term periods, such as hourly or 15 minutes, are useful by means of crossovers and by reaching short-term overbought/oversold areas. Other oscillators help as well, but you have to select them.

Those of you having access to the currency futures prices and charts can take advantage of price gaps, even if you trade spot (as opposed to currency futures). The percentage retracements may come into play following extended moves. The parabolic is most sensitive in the intraday trading, but still provides direction and stop-loss levels. Finally, some of you may look at the Market Profile® in an attempt to identify some significant direction early in the morning.

Technical Tools for Each Type of Market

Technical analysis offers a rich arsenal of weapons to use in various types of markets. Although you have to decide what suits you best, Figure 12.4 sorts out the general choices to ease your selection process. The types of market identified are *trading* and *dynamic*. A trading market may either range within a channel, with room for regular trading, or be flat, barely fluctuating in a limited trading range.

A dynamic market shows trends, trend continuation and reversal formations, and breakouts.

Trading Market

The trading market is the more common type. According to J. Peter Steidlmeyer's studies for his Market Profile®, you spend 80% of the time in trading markets and 5% in nontrending markets. (The other 15% is discussed in the next section.) You can use any type of chart or combination of charts, as long as you remember that the point and figure chart can generate whipsaws.

The support and resistance lines prove to be vital in giving you trading confidence. You may not be too happy trading in a less than exciting market, but the levels generated by the support and resistance lines bring you cash.

The stochastics' crossovers should help. This type of stochastics signal does not require extreme readings, nor does it signal major reversals. Since the crossovers can occur at any level, the opportunity for more signals increases. You can use both or either of the two types of the stochastics.

Other oscillators may provide further information. You must select the oscillators you consider to be most relevant to your currency and look for overbought/oversold signals.

| Trading | | Dynamic | |
Range or Flat	Trending	Reversal	Breakout
Line chart	Line chart	Line chart	Line chart
Bar chart	Bar chart	Bar chart	Bar chart
Candlestick chart	Candlestick chart	Candlestick chart	Point and figure chart
Point and figure chart	Point and figure chart	Point and figure chart	Candlestick chart
Support/resistance	Trend line	Classic trend reversal formations	Trend line
Stochastics crossover	Channel line	Candlestick reversal formations	Channel line
Oscillators OB/OS	Parabolic	Exhaustion gap, island, key reversal	Speedlines
Gann's angles	Classic trend continuation formations	Stochastics OB/OS	Support/resistance
Gann's Cardinal Square	Moving averages	Other oscillators OB/OS	Neckline
Gann's arcs	Measurement gap	Gann's Cardinal Square	Speedlines
Parabolic	Gann's Cardinal Square	Gann's angles	Gann's angles
Speedlines	Gann's angles	Gann's arcs	Gann's Cardinal Square
Candlestick "wait-and-see" formations	Elliott Wave	Moving averages	Breakaway gap
		Elliott Wave	Bollinger bands
		Tom DeMark	
		Gann's squaring	
		Gann's calendar	

Gann analysis offers you a multitude of technical signals. In a trading market, Gann's angles and, for that matter, the speedlines furnish you with firm levels. The Cardinal Square holds the most significant support/resistance levels, and the arcs identify the areas of potential reversal.

The parabolic system works best in trending markets, but you can make excellent use of it for short-term directions.

Finally, the Japanese "wait-and-see" studies should encourage you to be more disciplined and to do less trading while the market is making up its mind.

Dynamic Market

Trending

Following up on J. Peter Steidlmeyer's studies for his Market Profile®, you spend about 15% of your trading time in trending markets—and here is where you have the most fun. Although all the types of charts are good in a trending market, trend lines and channel lines provide you with the most substantial signals. The parabolic filters out trends, and the overall Gann analysis comes in handy as well. Moving averages work best in trending markets; so this is your opportunity to use them.

If you trade currency futures or have access to currency futures charts, the identification of a measurement gap gives you confidence in the trend, along with a clear price and duration target.

Reversal

You can identify reversals on all charts, although this works particularly well on line charts. Naturally, you receive the best signals from the classic and candlestick trend reversal formations. Exhaustion gaps, islands, and key reversals help those watching the currency futures charts.

Reversals signals should also be generated by some oscillators, especially the stochastics. Gann analysis, as usual, offers you a wide array of tools. Moving averages are slow to identify the reversal, but a divergence in excess of 25% between two consecutive averages warns you, while a golden square intersection confirms the currency reversal. The Bollinger bands can also yield reversal signals. Subject to confirmation by other studies, the extreme divergence of these bands forecasts a currency reversal.

Breakouts

You can use any type of chart to identify breakouts, but the line chart is best because it plots the most important price of the day, the closing price.

The penetration of the trend lines, channel lines, speedlines, Gann's angles, and/or necklines provides the most powerful breakout signals. The penetration of the support and resistance lines furnishes both very significant and less significant technical signals.

The breakaway gap, identifiable on currency futures charts, reflects the market's swing outside a trading range. Again, Gann's Cardinal Square provides you with significant support/resistance levels whose penetration may generate a currency breakout. Bollinger bands help you identify the periods of low volatility, thus warning you of the upcoming increase in volatility and the breakout from the trading range.

Typical Problems in Technical Analysis

The wealth of information and the large number of methods available may, at times, lead traders to mistakenly believe that technical analysis does all the work for them. Technical analysis, however, requires hard work, discipline, and decisive action. Employ the methods you have studied; do not make them up as you go along. Traders who put unwarranted weight on incomplete or unclear patterns should not be surprised when they realize losses.

Lack of Confirmation

As you recall from Chapter 2, a major trend reversal formation, such as the head and shoulders formation, comes into being only when the neckline is penetrated on high volume at the closing.

Therefore, in Figure 12.5 you have no confirmation at the following levels:

- At point A, which is the closing time, because the currency may move either way.
- At point B, which is the closing time, if the volume is light.
- At point C, because it occurs during the trading day—you don't know where it will close.

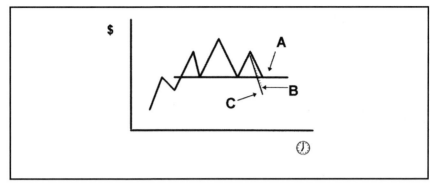

Figure 12.5. Lack of confirmation in a major trend reversal formation.

Anemic Breakouts of the Currency Through Trend and Channel Lines

Closely related to the lack of confirmation is the issue of anemic break-outs of the currency through either trend or channel lines (see Figure 12.6). Bar and candlestick charts generally reveal these breakouts in a time frame ranging between seconds to minutes. Naturally, you also see, at times, closings outside the trend and channel lines, but the lack of follow-through negates their breakout potential. In this situation, you have to adjust to the circumstances; perhaps you have to redraw the trend or channel lines, or maybe you should just ignore it. Again, for longer-term speculators and hedgers, the line chart is more reliable than bar or candlestick charts for channel penetration.

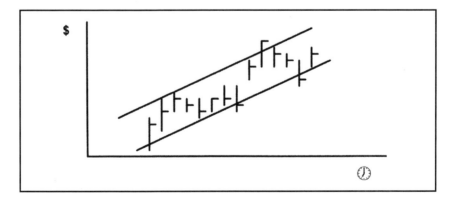

Figure 12.6. Anemic breakouts of the currency through either trend or channel lines.

Anemic Breakouts of the Currency Through Support and Resistance Lines

The anemic breakout of the support and resistance levels has the same technical impact as the anemic penetration of the trend and channel lines. All the lack of interest reflects is an isolated execution of a stop-loss order or a failed attempt to spearhead a new trading attack. A perfect example occurred in 1995, when the USD/JPY reached its record low (see Figure 12.7). As the overall market exhausted itself at the major support level of 80.00, a handful of traders in Asia sold the USD/JPY at 79.75. There was hardly any interest at those levels, and the currency bounced back aggressively.

Ambiguous Patterns

Traders love neat and reliable chart formations, but they must make sure that they do not love them to death. Every major chart pattern has certain characteristics. Figure 12.8 shows something like a head and shoulders formation. The first two rallies are roughly similar in height and the third one is meek. Even if this yet-to-be-named formation breaks below the quasineckline, you still cannot expect a trend reversal or a price objective.

Figure 12.7. An anemic break of major support in the US dollar/Japanese yen. (*Source:* Bridge Information Systems, Inc.)

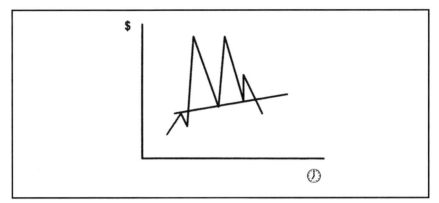

Figure 12.8. An ambiguous head and shoulders.

Unreasonable Price Objectives

Major chart formations that occur over the long term may be very large. Although they perform as expected, these formations may stop short of reaching the price objective. This happens because reaching these targets means taking the currency to unprecedented levels. This was the case of the triple tops in the British pound in September 1992 (Figure 12.2). The pound was marked to reach a low of about 1.3200, but that level was unreasonably low. The currency actually stopped—or was helped to stop—at around 1.4250. Although the neckline was broken and the currency was severely sold, it could not reach its unreasonable price objective.

Wrong Identification of a Pattern

A potential hazard may result from wrongfully identifying a pattern. Your cost may be a downright loss or a limited profit. An example is the identification of a triangle instead of a pennant. As you can see in Figure 12.9, the price target is very different: The target of the pennant is 1.4420, whereas the target of the triangle is only 1.4230.

Anemic Breakout of Oscillators Through the Zero Line

When an open scale oscillator, such as the momentum, crosses above the zero line, expect the currency to advance. When it falls below the zero line, expect to sell the currency. But this type of trading signal tells you only half the story. Equally important, the penetration must be relevant in terms of magnitude and time. As presented in Figure 12.10, the anemic crossover at point *A* has no technical importance.

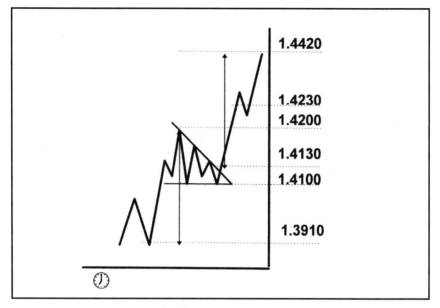

Figure 12.9. A wrong pattern identification.

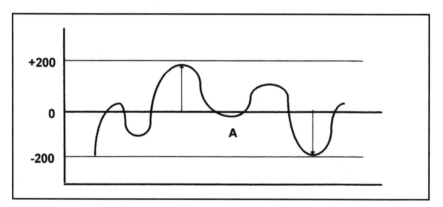

Figure 12.10. An anemic crossover of the zero line by an open scale oscillator.

Anemic Crossovers between Moving Averages or Oscillators

The intersection of two consecutive moving averages or two oscillators must also have magnitude and time significance. Simply disregard temporary crossovers of the type presented in Figure 12.11.

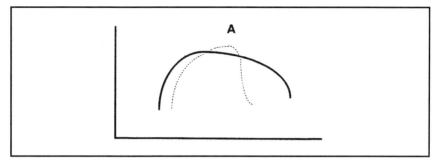

Figure 12.11. An anemic crossover of two consecutive moving averages or two oscillators.

Chart Combinations

Selecting the types of charts to use in your technical analysis is a very personal choice. However, since different types of charts may provide different action or timing signals, your choice must be based on whether you are a short-term or a long-term trader. Naturally, if you do not have access to the types of charts you want, do the best you can with the available chart or charts. Figures 12.12 and 12.13 illustrate potential chart combinations.

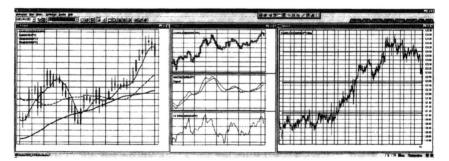

Figure 12.12. Potential chart combinations in euro/US dollar.
(*Source:* Bridge Information Systems, Inc.)

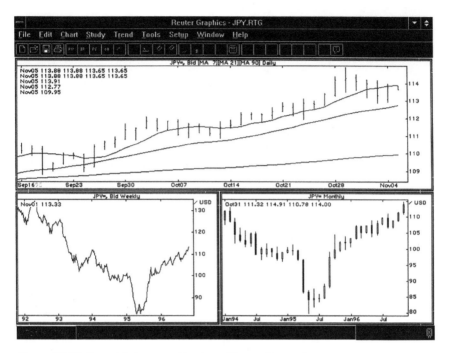

Figure 12.13. Potential chart combinations. (*Source:* Reuters.)

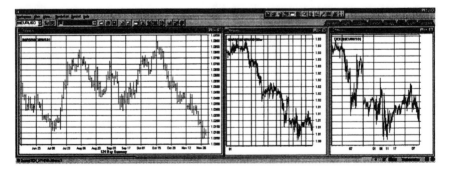

Figure 12.14. Potential chart combinations in euro/US dollar.
(*Source:* Bridge Information Systems, Inc.)

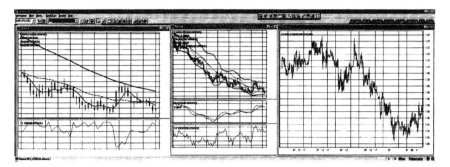

Figure 12.15. Potential chart combinations in euro/US dollar. (*Source:* Bridge Information Systems, Inc.)

Figures 12.14 and 12.15 illustrate potential chart combinations with studies.

Where to Go from Here

If you trade a portfolio of currencies that keeps you very busy, you can use alarms as a simple technological aid. Just leave certain price levels in your system, and the alarm faithfully rings when those levels are reached.

For example, if you want to save time, or if you want to test your knowledge of candlestick analysis, use the pattern recognition features of CQG (see Figure 12.16). Many software systems offer pattern recognition for Elliott Wave and Tom DeMark studies. This is especially helpful for the latter two, where the tedious manual analysis necessary discourages many technicians.

If you do not like the hassle of spending a lot of time analyzing charts, if you just lack the time, or if you want to objectively apply a subjectively chosen mix of technical tools, then you may want to go a step further and use an automatic trading system. Once programmed with the mixture of studies you consider to be reliable and profitable, the system does the analytical work, and you are left only with the execution. As the technology advances rapidly, this type of system is becoming more pop-

Figure 12.16. Example of candlestick pattern identification on a US dollar/Japanese yen chart. (*Source:* CQG. ©Copyright CQG INC.)

ular among individual traders than among interbank currency traders who tend to rely on their own analysis because they are generally blessed with few currencies to track. Automatic trading systems are very popular among equity technicians because they must analyze tens and hundreds of issues in short periods of time.

Beyond automatic trading systems is the realm of neural networking. These pattern recognition programs improve their performance by trial and error, and they are used as "black boxes." In other words, you do not need the mathematical calculations, just the results. One problem with the black box approach is that it requires a lot of statistical cleaning. Another problem is that black boxes at times identify insignificant patterns and then project them into a mainstream result. For instance, suppose a currency is in a major up trend, but, for whatever reason, there is some profit taking for three Wednesdays in a row. The system might tell you that every single Wednesday the currency will drop.

These sophisticated systems might be able to identify patterns, but they cannot think. Thinking is still a human prerogative. To overcome this rough approach, there are two solutions: genetic algorithms and fuzzy logic. *Genetic algorithms* are also based on the trial-and-error process used by black boxes, but the approach is analogous natural selection, as in the evolutionary process. *Fuzzy logic* introduces shades of gray to the black box's world of primary colors. The programmers instruct the software not only in clear positive and negative factors, but also in terms of "to some extent."

Efforts to quantify the correlation among multiple financial markets and thus forecast the impact of either main events or marginal events on your currency are commendable, and they can be successful in certain cases. However, there are problems.

One problem is poor asset allocation. For example, a player may allocate equal weight to the four major currencies when only one is trending. Obviously, the system should be reprogrammed to focus mainly, if not totally, on the trending currency.

Another problem is in trying to transfer the equity portfolio approach to the currency market. In this instance, a player opens a portfolio of about one hundred currencies to spread the risk. However, trading with any degree of efficiency outside the major or secondary currencies is not possible. The lack of liquidity outside the main currencies increases the overall risk, rather than lowers it.

Finally, we live in an increasingly complex world with very diverse needs, levels of understanding, and tools of implementation. Thus, even in the most efficient financial markets, such as the foreign exchange, efficiency goes only so far. Technology does not destroy the market by finding the "key" to its secret. The market is made up of people, and the purely human players in the market keep their "secrets" for as long as people are diverse. However, technology allows many of us to take our forecasting to levels unimaginable only decades ago.

Technical analysis is quantifiable, not that difficult to grasp, and very profitable. It is a splendid approach to forecasting, usable by diverse types of players in diverse time frames for diverse reasons. The purpose of this book is to arm you with all the weapons that the rest of the market uses only infrequently. Choose your best set, test it, challenge it, use it.

Your control of the technical forecasting tools turns you from a market participant into a successful and confident player. You will make more decisions than guesses, you will discover new paths, and you will find new approaches to profitability.

Good hunting!

BIBLIOGRAPHY

Frost, A. J. and Prechter Jr., Robert R., *Elliott Wave Principle,* 5th ed. New Classics Library, 1985.

Luca, Cornelius, *Trading in the Global Currency Markets,* 2nd ed. New York Institute of Finance, 1999.

Murphy, John J., *Technical Analysis of the Futures Market.* New York Institute of Finance, 1986.

Pring, Martin J., *Martin Pring on Market Momentum.* International Institute for Economic Research, Inc., 1993.

Shimizu, Seiki, *The Japanese Chart of Charts.* Tokyo Futures Trading Publishing Co., Tokyo, Japan, 1986.

GLOSSARY

Accumulation Swing Index (ASI) An oscillator based on the Swing Index (SI). A *buying signal* is generated when the daily high exceeds the previous SI significant high, and a *selling signal* occurs when the daily low dips under the significant SI low.

Akasansen shianboshi *See* Red three candlestick in deliberation.

Amplitude The height of a cycle.

Andrews' pitchfork A method of channel identification.

Ascending triangle A triangle continuation formation with a flat upper trend line and a bottom sloping upward trend line (*see* Triangles).

Ascending triple top A bullish point and figure chart formation suggesting that the currency is likely to break a resistance line the third time it reaches it. Each new top is higher than the previous one.

Ate See Atekubi.

Atekubi A bearish two-day candlestick combination. It consists of a blank bar that closes at the daily high; the current closing price equals the previous day's low. The original day's range is a long black bar.

Average directional movement index (ADX) A version of the directional movement index to be used on periods of high volatility.

Bar chart A type of chart consisting of four significant points: the *high* and the *low* prices, which form the vertical bar; the *opening* price, which is marked with a little horizontal line to the left of the bar; and the *closing* price, which is marked with a little horizontal line to the right of the bar.

Bearish divergence A chart event in which the currency moves up while the oscillator moves down.

Bearish *tasuki* A bearish two-day candlestick combination. It consists of a long blank bar that has a low above 50% of the previous day's long black body and that closes marginally above the previous day's high. The second day's rally is temporary, because it is caused only by profit taking. The selloff is likely to continue the next day.

Bearish *tsutsumi* (the engulfing pattern) A bearish two-day candlestick combination. It consists of a second-day long black candlestick whose body "engulfs" the previous day's small blank body.

Black closing *bozu* A bearish candlestick formation that consists of a long black bar (upper shadow).

Black *marubozu* (shaven head) A bearish candlestick formation that consists of a long black bar (no shadow).

Black opening *bozu* A bearish candlestick formation that consists of a long black bar (lower shadow).

Blank closing *bozu* A bullish candlestick formation that consists of a long blank bar (lower shadow).

Blank *marubozu* (shaven head) A bullish candlestick formation that consists of a long blank bar (no shadows).

Blank opening *bozu* A bullish candlestick formation that consists of a long blank bar (upper shadow).

Body of candlestick The difference between the high and the low of the period encompassed by the candlestick (*see Jittai*).

Bollinger bands A quantitative method that combines a moving average with the instrument's volatility. The bands were designed to gauge whether the prices are relatively high or low. They are plotted two standard deviations above and below a simple moving average. The bands look like an expanding and contracting envelope model. When the band contracts drastically, the signal is that volatility will expand sharply in the near future. An additional signal is a succession of two top formations: one outside the band, followed by one inside. If it occurs above the band, it is a selling signal. When it occurs below the band, it is a buying signal.

Book method The original name of the point and figure chart.

Breakaway gap A price gap that occurs in the beginning of a new trend, many times at the end of a long consolidation period. It may also appear after the completion of major chart formations.

Breakout of a spread triple bottom A bearish point and figure chart formation suggesting that the currency is likely to break a support line the third time it reaches it. The currency fails to reach the support line once.

Breakout of a spread triple top A bullish point and figure chart formation suggesting that the currency is likely to break a resistance line the third time it reaches it. The currency fails to reach the resistance line once.

Breakout of a triple bottom A bearish point and figure chart formation suggesting that the currency is likely to break a support line the third time it reaches it.

Breakout of a triple top A bullish point and figure chart formation suggesting that the currency is likely to break a resistance line the third time it reaches it.

Bullish divergence Chart event in which the currency moves down while the oscillator moves up.

Bullish *tasuki* A bullish two-day candlestick combination. It consists of a long black bar that has a high above 50% of the previous day's long blank body and closes marginally below the previous day's low.

Candlestick chart A type of chart that consists of four major prices: high, low, open, and close. The *body (jittai)* of the candlestick bar is formed by the opening and closing prices. To indicate that the opening is lower than the closing, the body of the bar is left blank. If the currency closes below its opening, the body is filled. The rest of the range is marked by two "shadows": the *upper shadow (uwakage)* and the *lower shadow (shitakage)*.

Cardinal Square A Gann technique of forecasting future significant chart points by counting from the all-time low price of the currency. It consists of a square divided by a cross into four quadrants. The all-time low price is housed in the center of the cross. Each of the following higher prices is entered in clockwise order. The numbers positioned in the cardinal cross are the most significant chart points.

Channel line A parallel line that can be traced against the trend line, connecting the significant peaks in an up trend and the significant troughs in a down trend.

Commodity channel index (CCI) An oscillator that consists of the difference between the mean price of the currency and the average of the mean price over a predetermined period of time.

Commodity Research Bureau's (CRB) Futures Index Index formed from the equally weighted futures prices of 21 commodities.

Common gap A price gap that occurs in relatively quiet periods or in illiquid markets. It has a limited technical significance.

Continuation patterns Technical signals that reinforce the current trends.

Currency call option A contract between the buyer and seller holding that the buyer has the right, but not the obligation, to buy a specific quantity of a currency at a predetermined price and within a predetermined period of time, regardless of the market price of the currency. The *writer* assumes the obligation of delivering the specific quantity of a currency at a predetermined price and within a predetermined period of time, regardless of the market price of the currency, if the buyer wants to exercise the call option.

Currency futures A specific type of forward outright that deals with the expiration date, the size of the amount, and the methods of trading.

Currency option *See* Currency call option; Currency put option.

Currency put option A contract between the buyer and the seller holding that the buyer has the right, but not the obligation, to sell a specific quantity of a currency at a predetermined price and within a predetermined period of time, regardless of the market price of the currency. The *writer* assumes the obligation of buying the specific quantity of a currency at a predetermined price and within a predetermined period of time, regardless of the market price of the currency, if the buyer wants to exercise the call option.

Cycles A propensity for events to repeat at about the same time.

Dead cross An intersection of two consecutive moving averages that move in opposite directions and that, technically, should be disregarded.

Descending triangle A triangle continuation formation with a flat lower trend line and a downward sloping upper trend line (*see* Triangles).

Descending triple bottom A bearish point and figure chart formation suggesting that the currency is likely to break a support line the third time it reaches it. Each new bottom is lower than the previous one.

Diamond A minor reversal pattern that resembles a diamond.

Directional movement The part of the current day's range falling outside yesterday's range.

Directional movement index (DMI) A signal of trend presence in the market. The line simply rates the price directional movement on a scale of 0 to 100. The higher the number, the better the trend potential of a movement, and vice versa.

Double bottoms A bullish reversal pattern that consists of two bottoms of approximately equal heights. A parallel (resistance) line is drawn against a line that connects the two tops. The break of the resistance line generates a move equal in size with the price difference between the average height of the bottoms and the resistance line.

Double tops A bearish reversal pattern that consists of two tops of approximately equal heights. A parallel (support) line is drawn against a resistance line that connects the two tops. The break of the support line generates a move equal in size with the price difference between the average height of the tops and the support line.

Down gap three wings Version of the bearish *sanpei*.

Downside *tasuki* gap A bearish two-day candlestick combination. It consists of a second-day blank bar that closes an overnight gap opened on the previous day by a black bar.

Dow Theory Technical analysis theory developed on the study of the stock price movement, which identifies the reaction of the market to events, defines and sorts the trends, and gauges the worthiness of the trends.

Downward breakout from a consolidation formation A bearish point and figure chart formation that resembles the inverse flag formation. A valid downside breakout from the consolidation formation has a price target equal in size to the length of the previous down trend.

Downward breakout of a bearish support line A bearish point and figure chart formation that confirms the currency's breakout of a support line the third time it reaches it.

Downward breakout of a bullish support line A bearish point and figure chart formation that confirms the currency's breakout of a support line the third time it reaches it. The support line is sloped upward.

Elliott Wave principle A system of empirically derived rules for interpreting action in the markets. It refers to a five-wave/three-wave pattern that forms one complete bull market/bear market cycle of eight waves.

Engulfing pattern Second-day candlestick that engulfs and closes outside the previous day's range (*see Tsutsumi*).

Envelope model A band created by two winding parallel lines above and below a short-term moving average will create a band bordering most price fluctuations. When the upper band is penetrated, a *selling signal* occurs. When the lower band is penetrated, a *buying signal* is generated. Since the signals generated by the envelope model are very short-term geared and they occur many times against the ongoing direction of the market, speed of execution is paramount.

Evening southern cross Version of the three river evening star pattern.

Exhaustion gap Price gap that occurs at the top or at the bottom of a V-formation. The trend changes direction in a rather uncharacteristically quick manner.

Expanding (broadening) triangle A triangle continuation formation that looks like a horizontal mirror image of a triangle, where the tip of the triangle is next to the original trend, rather than its base (*see Triangles*).

Exponentially smoothed moving average A moving average that also takes into account the previous price information of the underlying currency.

Fibonacci arcs Method of price and time support and resistance identification based on the Fibonacci ratio.

Fibonacci fans Method of price support and resistance identification based on the Fibonacci ratio.

Fibonacci percentage retracements Price retracements of .382 and .618, or approximately 38% and 62%.

Fibonacci ratio 0.312, 0.618, 1.312, and 1.618.

Fibonacci sequence Takes a sequence of numbers that begins with 1 and adds 1 to it, takes the sum of this operation and adds it to the previous term in the sequence. Then it takes the sum of this second operation and adds it to the previous term in the sequence. The Fibonacci sequence continues iterating in this manner, adding the most recent sum to the previous term, which is itself the sum of the two previous terms, etc. This yields the following series of numbers: 1, 1, 2, 3, 5, 8, 13, 21, 34, 55, 89, 144, 233, 377, (etc.).

FINEX A currency market, part of the New York Cotton Exchange (NYCE), the oldest futures exchange in New York. The exchange lists futures on the euro and the USDX.

Flags A continuation formation that resembles the outline of a flag. It consists of a brief consolidation period within a solid and steep upward or downward trend. The consolidation itself tends to be sloped in the opposite direction from the slope of the original trend, or simply flat. The consolidation is bordered by a support line and a resistance line, which are parallel to each other or very mildly converging, making it look like a flag (parallelogram). The previous sharp trend is known as the flagpole. Once the currency resumes its original trend by breaking out of the consolidation, the price objective is the total length of the flagpole, measured from the breakout price level.

Forward outright Foreign exchange deal that matures a day past the spot delivery date (generally two business days).

Forward spread (forward points or forward pips) Forward price used to adjust a spot price to calculate a forward price. It is based on the current spot exchange rate, interest rate differential, and the number of days to delivery.

Four-week rule Richard Donchian's rule suggesting to go long when the currency exceeds the highs of the previous four trading weeks, and to sell when the currency falls below the lowest price of the previous four business weeks.

Fractal geometry Geometry theory that refers to the fact that certain irregular objects have a fractal number of dimensions. In other words, an object cannot fill an integer number of dimensions, i.e., a spec of dust has a real dimension of between 0 and 1.

Fuzzy logic Method that attempts to weigh the quality of the patterns recognized by neural networks. Since not all patterns have equal financial significance for foreign currency forecasting, this method qualifies the degree of certainty of the results.

Gann percentage retracements The Gann theory focuses mostly on the 3/8, 4/8 and 5/8, or 38%, 50% and 62% retracement figures.

Gap The price gap between consecutive trading ranges (i.e., the low of the current range is higher than the high of the previous range).

Genetic algorithms The method used to optimize a neural network. Trial and error is applied to an evolution-like system, which generates natural selection for financial forecasting purposes.

Golden cross An intersection of two consecutive moving averages that move in the same direction and suggests that the currency will move in the same direction.

Hammer A small body candlestick with a very long tail that acts as a bear trap (*see Karakasa*).

Hangman A small body candlestick with a very long tail that acts as a bull trap (*see Karakasa*).

Harami A "wait-and-see" two-day candlestick combination. It consists of two consecutive ranges having opposite directions, but it does not matter which one is first. The second day's range falls within the previous day's body.

Haramiyose A version of *harami* where the second day is a *doji* candlestick.

Head and shoulders A bearish reversal pattern that consists of a series of three consecutive rallies, where the first and third rallies (the shoulders) have about the same height and the middle one (the head) is the highest. The rallies are based on the same support line, known as the neckline. When the neckline is broken, the price target is approximately equal in amplitude to the distance between the top of the head and the neckline.

High-low band A band created by two winding parallel lines above and below a short-term moving average creates a band bordering most price fluctuations. The moving average is based on the high and low prices. The resulting two moving averages define the edges of the band. A close above the upper band suggests a buying signal, and one below the lower band gives a selling signal.

Hinge A slowdown in the velocity of the %K line and/or the %D line in the stochastics, which is a reversal warning.

Horizontal count A method used in point and figure charting to extrapolate vertically the number of columns in a fulcrum on top of the breakout price.

Hoshi See Star.

In gear A chart event in which the currency and the oscillator move in the same direction.

Inside day A daily range narrower than yesterday's range.

Inverted head and shoulders A bullish reversal pattern that consists of a series of three consecutive selloffs. Among the selloffs, the shoulders have approximately the same amplitude, and the head is the lowest. The formation is based on a resistance line called the neckline. After the neckline is penetrated, the target is approximately equal in amplitude to the distance between the top of the head and the neckline.

Irikubi A bearish two-day candlestick combination. It consists of a modified *atekubi* bar. All the characteristics are the same, except that the second day's closing high is marginally higher than the original day's low.

Island reversal An isolated range or ranges that occur at the tip of a V-formation.

Jittai Body of the candlestick.

Kabuse (**dark cloud cover**) A bearish two-day candlestick combination. It consists of a second-day long black bar that opens above the high of the previous day's blank bar and closes within the previous day's range (in an up trend).

Karakasa (*hangman* at the top, *hammer* at the bottom) A bearish candlestick at the top of the trend, bullish at the bottom of the trend. The candlestick can be either blank or black. The body of the candlestick is very small and only half the length of the shadow.

Keltner channels A method of minor trend identification.

Kenuki See Tweezers.

Kenukitenjo Top tweezers.

Kenukiyosesen A version of the top tweezers where the second day is a *doji* candlestick.

Key reversal day The daily price range on the bar chart of the reversal day fully engulfs the previous day's range, and also the close is outside the preceding day's range.

Kirikaeshi See Kirikomi.

Kirikomi (*kirikaeshi*) A bullish two-day candlestick combination. It consists of a blank marubozu bar that opens the second day lower than the previous low of a long black line and closes above the 50% level of the previous day's range.

Koma (**spinning tops**) A reversal candlestick formation that consists of a short bar, either blank or black. This candlestick may also suggest lack of direction.

Linearly weighted moving average A moving average that assigns more weight to the more recent closings.

Line chart The line connecting single prices for each of the time periods selected.

Logarithmic scale A method of chart plotting that takes into consideration the percentage price change.

Long legged shadows' *doji* A reversal candlestick formation that consists of a bar in which the opening and closing prices are equal.

Long straddle A compound option that consists of a long call and a long put on the same currency, at the same strike price and with the same expiration dates. The maximum loss for the buyer is the sum of the premiums. The up side break-even point is the sum of the strike price and the premium on the straddle. The down side break-even point is the difference between the strike price and the premium on the straddle. The loss is unlimited.

Long strangle A compound option that consists of a long call and a long put on the same currency, at the different strike prices, but with the same expiration dates. The loss is unlimited.

Lower shadow The difference between the low and closing prices on a bearish day, or the difference between the low and the opening prices on a bullish day in candlestick charting (*see Shitakage*).

Market profile An indicator designed to analyze how much trading activity takes place in the futures market, at what price, and at what time.

Momentum An oscillator designed to measure the rate of price change, not the actual price level. This oscillator consists of the net difference between the current closing price and the oldest closing price from a predetermined period. The momentum is measured on an open scale around the zero line.

Moving average An average of a predetermined number of prices over a number of days, divided by the number of entries.

Moving average convergence/divergence (MACD) An oscillator that consists of two exponential moving averages (other inputs may be chosen by the trader as well) that are plotted against the zero line. The zero line represents the times the values of the two moving averages are identical. The *buying signal* is generated when this intersection is upwards, whereas the *selling signal* occurs when the intersection takes place on the down side.

Moving averages oscillator An oscillator in which the values of two consecutive moving averages are subtracted from each other (the larger number of days from the previous one) and the new values are plotted.

Neural networks Computer systems that recognize patterns. They may be used to generate trading signals or to be part of trading systems.

On balance volume Indicator designed to gauge the soundness of a trend, which can be used in the currency futures market.

Oscillators Quantitative methods designed to provide signals regarding the overbought and oversold conditions.

Outside day Daily range wider than yesterday's range.

Parabolic system (SAR) A stop-loss technical system, based on price and time. *SAR* stands for stop and reverse. The stop moves daily in the direction of the new trend. The built-in acceleration factor pushes the SAR to catch up with the currency price. If the new trend fails, the SAR signal is generated.

The name of the system is derived from its parabolic shape, which follows the price gyrations. It is represented by a dotted line. When the parabola is placed under the price, it suggests a long position, and an above-the-price parabola indicates a short position.

Pennant A continuation formation that resembles the outline of a pennant. It consists of a brief consolidation period within a solid and steep upward or downward trend. The consolidation itself tends to be sloped in the opposite direction from the slope of the original trend, or simply flat. The consolidation is bordered by a support line and a resistance line, which converge, creating a triangle. The previous sharp trend is known as the pennant pole. Once the currency resumes its original trend by breaking out of the consolidation, the price objective is the total length of the pole, measured from the breakout price level.

Period The length of a cycle.

Phase The current location on a cycle wave.

Point and figure chart A type of chart that plots price activity without regard to time. When the currency moves up, the fluctuations are marked with Xs. The moves on the down side are plotted with Os. The direction on the chart changes only if the currency reverses by a certain amount of pips.

Random Walk Theory, The An efficient market hypothesis, stating that prices move randomly versus their intrinsic value. Therefore, no one can forecast market activity based on the available information.

Range The difference between the high and the low prices for a specific period.

Range expansion index A method of identifying trend patterns and trading signals.

Rate of change (ROC) A momentum oscillator in which the oldest closing price is divided into the most recent one.

Rectangle A continuation formation that resembles the outline of a parallelogram. The price objective is the height of the rectangle.

Red three candlestick advance block A version of the three parallel candlesticks.

Red three candlestick in deliberation A version of the three parallel candlesticks.

Relative strength index An oscillator that measures the relative changes between the higher and lower closing prices. The RSI is plotted on a 0 to 100 scale. The 70 and 30 values are used as warning signals, whereas values above 85 indicate an overbought condition (selling signal), and, under 15, an oversold condition (buying signal).

Resistance level The peaks representing the price level where the supply exceeds the demand.

Reversal patterns Patterns that occur at the end of the trend, signaling trend change.

Rickshaw man A candlestick with no body, which occurs in a midrange.

Rounded bottom A bullish reversal pattern that consists of a very slow and gradual change in the direction of the market.

Rounded top (saucer) A bearish reversal pattern that consists of a very slow and gradual change in the direction of the market.

Runaway or measurement gap A price gap that occurs within solid trends. It is also called a measurement gap because it tends to occur about midway through the life of a trend.

Sakizumari See Red three candlestick advance block.

Sangu **(three gaps)** A complex reversal candlestick method applicable in either a steeply rising or falling market, where the time limits break the trading. After the third gap occurs, the market should reverse at least to fill the second gap.

Sanpei **(three parallel bars)** A complex reversal candlestick combination that refers to the similarity of direction and velocity of three consecutive bars. When bullish, the formation is known as the three soldiers. When bearish, the name is the three crows.

Sanpo **(three methods)** A complex candlestick combination advising that retracements are in order before the market can reach new highs or new lows.

Sansen **(three rivers) method** A complex reversal candlestick combination similar to an exhaustion gap. It consists of three candlesticks, where the middle one is short and either above, or below, the other two.

Sanzan **(three mountains)** A complex reversal candlestick combination that consists of a triple top formation.

SAR *See* Parabolic system (SAR).

Sashikomi A bearish two-day candlestick combination that consists of a modified *irikubi* bar. The difference is that the opening of the second day's blank bar is much lower than that of the *irikubi* bars. Despite the wider gap thus formed, the blank candlestick closes just above the previous day's low.

Sequential An indicator designed to identify the areas of exhaustion.

Shitakage A lower shadow of the candlestick (*see* Candlestick chart).

Shooting star Version of the star pattern in which the shadow of the star is long.

Simple moving average, or arithmetic mean An average of a predetermined number of prices over a number of days, divided by the number of entries.

Simultaneous three wings Version of the bearish *sanpei.*

Slow stochastics A version of the original stochastic oscillator. The new slow %K line consists of the original %D line. The new slow %D line formula is calculated off the new %K line.

Speedlines Support or resistance lines that divide the range of the trend into thirds on a vertical line. The two resulting speedlines are plotted using as coordinates the origin and the 1/3 and 2/3 prices, respectively.

Spot deal A foreign exchange deal that consists of a bilateral contract between a party delivering a certain amount of a currency and a party receiving an amount of another currency, based on an agreed exchange rate, within two business days of the deal date. The exception is the Canadian dollar, in which the spot delivery is executed within one business day.

Spring A bear trap in the bar chart leading to a bullish reversal.

Star A "wait-and-see" two-day candlestick combination. It consists of a tiny body that appears the following day outside the original body. It is not important whether the star reaches the previous day's shadows. The direction of the two consecutive candlesticks is irrelevant.

Stochastics An oscillator that consists of two lines called %K and %D. Visualize %K as the plotted instrument and %D as its moving average. The resulting lines are plotted on a 1 to 100 scale. Just as in the case of the RSI, the 70% and 30% values are used as warning signals. The *buying (bullish reversal) signals* occur under 10% and the *selling (bearish reversal) signals* come into play above 90%.

Support level The troughs representing the level where the demand exceeds the supply.

Swap deal A foreign exchange deal that consists of a spot deal and a forward outright deal. A party simultaneously buys and sells (or sells and buys) the same amount of a currency with another counterparty, where the two legs of the transaction mature on different dates (one of the dates being the spot date) and are traded at different exchange rates (one of the exchange rates being the spot rate).

Swing index (SI) A momentum oscillator that is plotted on a scale of −100 to +100. The spikes reaching the extremes suggest reversal.

Symmetrical triangle A triangle continuation formation in which the support and resistance lines are symmetrical (*see* Triangles).

TD arcs A method of retracement analysis for both price and time.

Technical analysis The chart study of past behavior of commodity prices for forecasting their future performance.

Three Buddha reversal formation A complex reversal candlestick combination. It consists of a head and shoulders formation, or three consecutive rallies in which the first and the third are of approximately the same height, and the second is the highest.

Three crows Version of the bearish *sanpei*.

Three gaps *See Sangu.*

Three methods *See Sanpo.*

Three parallel candlesticks *See Sanpei.*

Three river evening star *See Sansen* (three rivers) method.

Three river morning star An inverted three river evening star pattern.

Three soldiers A version of the three parallel candlesticks.

Tick chart An intraday line chart that shows every transaction.

Tohbu (**gravestone *doji***) A reversal candlestick formation.

Tonbo (**dragonfly**) A reversal candlestick formation.

Traditional (Charles Dow) percentage retracements 1/3, 1/2 and 2/3, or 33%, 50% and 66%.

Trend The general direction of the market, as shown by the significant peaks and troughs of the currency fluctuations.

Trend line A straight line connecting the significant highs (peaks) in a down trend, and the significant lows (troughs) in an up trend.

Triangles A continuation formation that resembles the outline of a pennant, but without the pole. It consists of a brief consolidation period within a solid and steep upward trend or downward trend. The consolidation itself tends to be sloped in the opposite direction from the slope of the original trend, or simply flat. The consolidation is bordered by converging support and resistance lines, making it look like a triangle. Once the currency resumes its original trend by breaking out of the consolidation, the price objective is the height of the triangle, measured from the breakout price level.

Triple bottoms A bullish reversal pattern that consists of three bottoms of approximately equal heights. A parallel resistance line is drawn against a line that connects these tops. The break of the resistance line generates a move equal in size with the price difference between the average height of the bottoms and the resistance line.

Triple tops A bearish reversal pattern that consists of three tops of approximately equal heights. A parallel support line is drawn against a resistance line that connects these tops. The break of the support line generates a move equal in size with the price difference between the average height of the tops and the support line.

TRIX Index An oscillator that consists of one-day ROC calculation of a triple exponentially smoothed moving average of the closing price.

True range The greatest of the following: today's range, the difference between today's high and yesterday's closing, and the difference between today's low and yesterday's closing.

Tsutsumi *See* Engulfing pattern.

Tweezers A "wait-and-see" two-day candlestick combination. It consists of consecutive bars that have matching highs or lows. In a rising market, a tweezers top occurs when the highs match. The opposite is true for a tweezers bottom.

Two crows reversal A version of the three river evening star pattern.

Upper shadow The difference between the high and closing price on a bullish day, or the difference between the high and the opening price on a bearish day in candlestick charting (*see Uwakage*).

Upside gap *tasuki* A bullish two-day candlestick combination. It consists of a second-day black bar that closes an overnight gap opened on the previous day by a blank bar.

Upside gap's two crows A version of the three river evening star pattern.

Upthrust A bull trap in the bar chart leading to a bearish reversal.

Upward breakout from a consolidation formation A bullish point and figure chart formation that resembles the flag formation. A valid upside breakout from the consolidation formation has a price target equal in size to the length of the previous up trend.

Upward breakout of a bearish resistance line A bullish point and figure chart formation that confirms the currency's breakout of a resistance line the third time it reaches it. The resistance line is sloped downward.

Upward breakout of a bullish resistance line A bullish point and figure chart formation that confirms the currency's breakout of a resistance line the third time it reaches it.

USDX A currency index that consists of the weighted average of the prices of major foreign currencies against the US dollar.

Uwakage The upper shadow of the candlestick.

V-formations (spikes) A reversal formation that shows sudden trend changes and is accompanied by a heavy trading volume. This pattern may include a key reversal day or an island reversal and an exhaustion gap.

Volume The total amount of currency traded within a period of time, usually one day.

Volume accumulation oscillator (VAO) A more sensitive indicator of the soundness of a trend, which can be used in the currency futures market.

Wedges A continuation formation that resembles the outline of a pennant, but without the pole. It consists of a brief consolidation period within a solid and steep upward or downward trend. The consolidation is sharply angled in the opposite direction from the slope of the original trend. The consolidation is bordered by a support line and a resistance line that converge, making it look like a sharply angled triangle. Once the currency resumes its original trend by breaking out of the consolidation, the price objective is the height of the wedge, measured from the breakout price level.

Williams %R Version of the stochastics oscillator. It consists of the difference between the high price of a predetermined number of days and the current closing price, and the difference in turn is divided by the total range. This oscillator is plotted on a reversed 0 to 100 scale. Therefore the bullish reversal signals occur under 80%, and the bearish signals will appear above 20%. The interpretations are similar to those discussed under stochastics.

INDEX